# PERFLUORINATED POLYMER ELECTROLYTE MEMBRANES FOR FUEL CELLS

# PERFLUORINATED POLYMER ELECTROLYTE MEMBRANES FOR FUEL CELLS

TATSUHIRO OKADA, MORIHIRO SAITO
AND KIKUKO HAYAMIZU

Nova Science Publishers, Inc.

*New York*

**NOTICE TO THE READER**



**LIBRARY OF CONGRESS CATALOGING-IN-PUBLICATION DATA**

Perfluorinated polymer electrolyte membranes for fuel cells / Tatsuhiro Okada, Morihiro Saito, and Kikuko Hayamizu, editors.
    p. cm.
  ISBN 978-1-60456-804-2 (softcover)
  1. Conducting polymers. 2. Polyelectrolytes. 3. Organofluorine compounds. 4. Fuel cells. I. Okada, Tatsuhiro. II. Saito, Morihiro. III. Hayamizu, Kikuko.
  QD382.C66P47 2008
  621.31'2429--dc22

                    2008023214

*Published by Nova Science Publishers, Inc.* ≃ *New York*

# CONTENTS

# PREFACE

In this book the authors focus on the ion and water transport characteristics in Nafion® and other perfluorinated ionomer membranes that are recently attracting attention in various fields such as water electrolysis, mineral recovery, electrochemical devises and energy conversion. Methodology of measurements and data analysis are first presented that enable basic characterization of transport parameters in the perfluorinated ionomer membranes. Cation exchange isotherm data are collected in binary cation systems, with the aim to see the behaviors of cationic species that exist with $H^+$ in the membrane. Water transference coefficients, ionic transference numbers, ionic mobilities and other membrane transport parameters are measured in single and mixed counter cation systems using electrochemical methods. Diffusion coefficients of water and cations are also measured by pulsed-field-gradient spin-echo NMR (PGSE-NMR) at various temperatures in different kinds of perfluorinated ionomer membranes. The results are discussed in two perspectives.

One is to predict the hydration state in perfluorosulfonated ionomer membranes in relation to the possible degradation of performances in fuel cells under contaminated conditions with foreign cations. An analytical formulation of membrane transport equations with proper boundary conditions is proposed, and using various parameters of membrane transport, a simple diagnosis of water dehydration problem is carried out. This analysis leads one to an effective control of fuel cell operation conditions, especially from viewpoint of proper water management.

The others are to elucidate the ion and water transport mechanisms in the membrane in relation to polymer structures (e.g., different ion exchange capacity), and to propose a new design concept of polymer electrolyte membranes for fuel cell applications. Additionally for this purpose methanol and other alcohols are penetrated into the membrane, and alcohol permeability, membrane swelling, ionic conductivity and diffusion coefficients of water and $CH_3$ are measured systematically for various kinds of membranes to cope with the problem of methanol crossover in direct methanol fuel cells (DMFCs).

It is found that in order to realize a high ionic conductivity in the membrane, one should aim at a polymer structure through molecular design that takes into account the relative size of ions with a hydration shell against the size and atmosphere of ionic channels. For DMFC, a partially cross-linked polymer chain with high degree of hydrophilic ion transport paths based on phase-separated structures is recommended. Various possibilities of such polymer electrolytes are discussed.

# Chapter 1

# INTRODUCTION

Perfluorosulfonated ionomer (PFSI) membranes have been applied in a variety of fields such as chlor-alkali electrolysis, mineral extraction, synthesis, electrochemical sensors, and energy conversion systems [1-3]. Especially in the last decade their roles have increased as polymer electrolyte membranes applied in low temperature fuel cells (polymer electrolyte fuel cells, PEFC) [4-7]. Polymer electrolyte membranes are recognized as key materials that enable a very thin and strong layer of electrochemical cell, and therefore a high energy density in PEFC, owing to high proton conductivity, high mechanical, thermal, chemical and oxidative stabilities. Figure 1 shows chemical structures of some commercialized PFSI membranes. High concentration of sulfonic acid groups as cation exchange sites and high mobility of protons are specific features of such PFSI membranes. The polymer structure is characterized by hydrophilic ion cluster regions separated from hydrophobic main chains, forming a microscopic phase separation structure without cross-linking [8].

In order to improve performances of PEFC, it is essential to attain better characteristics of polymer electrolyte membranes, especially higher $H^+$ conductivities in the membrane. This will be achieved by complete understanding of mechanisms of $H^+$ transport kinetics in the ionic channel structures. It is known that the $H^+$ conductivity depends strongly on water content in the membrane [9]. Therefore, elucidating $H^+$ transport behaviors surrounded by water and local ion channel environment appears to be the main approach towards the goal of the optimum membrane materials in PEFC.

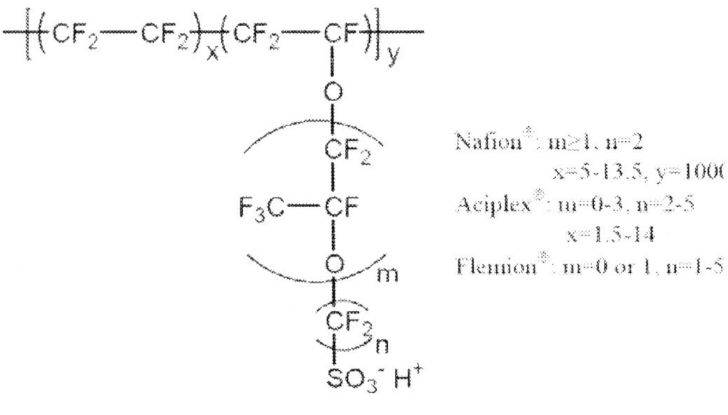

Figure 1. Chemical structures of Nafion®, Aciplex® and Flemion® membranes in H-from (with permission from the American Chemical Society).

To this end, both water and ion transport characteristics will be discussed first with relevance to the microscopic structures of PFSI membranes. This chapter starts with introduction of some morphological models of ion pair aggregation and cluster formation, together with reported experimental results such as diffraction or spectroscopic studies. Then transport models in the ionomer membranes are discussed that can fit the experimental data. New data are obtained from methodologies that are based on coupled linear transport equations in the membrane phase, paying attention to these models. This approach enables us to understand both water and ion transports in a mixed carrier system.

The mixed cation systems provide much information on the ionic interactions in the polymer membrane, from which the ion transport mechanism and mobility tendencies are derived based on some known parameters. Also the interaction of ion and water molecules inside the ionic channel proves to be a very important topic, which is of particular use in water management during PEFC operations. All the knowledge makes it possible to draw a picture of polymer structure and ion conduction, which is the ultimate goal in designing high performance $H^+$ conducting polymer membranes.

The importance of the mixed cation systems should be stressed more from practical implications. Normal membranes for applications may be used in H-form, but in special cases other cations may enter into the membrane, which is either made intentionally or caused by foreign impurity ions. In the chlor-alkali processes, Na-form membranes are used, and this system together with other cation systems is the main topic of investigations. In PEFC system, foreign ions possibly penetrate into the cell during the operation, either from the air stream at the cathode or from dissolved impurities through corrosion of materials. The effect of these foreign ions is unknown, and this problem should be clarified to avoid the degradation of PEFC. This makes also a major motivation of the present research, in order to keep the PEFC performances at the best conditions.

This chapter includes especially transport characteristics of perfluorinated ionomer membranes. To ease the understanding of such characteristics of the ionomer membranes, the structural and transport models are shortly reviewed historically, but the main topic is the transport parameters based on our recent experimental results. About the structural and morphological studies of the ionomers as well as thermal, mechanical and electrochemical behaviors, more than several hundreds research papers have been published, and the reader may refer to a wealth of monographs and recent review articles related to PEFC membranes [2,10-15].

*Chapter 2*

# STRUCTURE MODELS OF IONOMERS

## 2.1. SPECIFIC STRUCTURE OF PERFLUORINATED IONOMERS

"Ionomer" is defined as the category of ion-containing (or ion-conducting) polymers where the polymer structure is maintained by intermolecular forces in crystalline domains of main chains and ion-conducting paths form as the second domains [14]. This feature is different from ordinary ion exchange resins in which the structure is made of cross-linked polymer networks [16]. Although the solid-state ionomers have lower ion-exchange capacities than the ion exchange resins with structure of covalent cross-links, the former show higher ionic conductivity than the latter (table 1). The schematic picture of polymer structures in cross-linked and non cross-linked polymer electrolytes is presented in figure 2 [11]. It is pointed out that during the fabrication of non cross-linked proton conducting polymer, the "inverse micelle structure" is spontaneously formed, which consists of hydrophobic/hydrophilic phase separated structures. The aggregation of ion exchange sites makes the hydrophilic domain, and through this domain the ion conducts by the driving force of electric field. The amount of ion exchange groups is less than 15% of the mass of polymer repeating unit [13]. The degree of crystallinity as measured by diffraction analyses ranges between 12 and 22% [14].

We have several questions: how do these ionomers aggregate without cross-linking, what is the mechanism of generating specific hydrophilic domains, what is the details of hydrated structure around ion exchange groups and counter-ions and how to explain the formed size of ion cluster domains, etc., and they are the main themes to be solved. A structure model of ionomers should answer these questions starting with the assumed individual ion-pair forces between ion exchange sites and counter-ions as well as intermolecular forces among polymer main chains [13].

**Table 1. Comparison of Nafion® 117 and CR61 AZL 386 cation exchange membranes**

|  | Nafion® 117 | CR61 AZL 386 |
|---|---|---|
| **Producer** | DuPont | Ionics Inc. |
| **Membrane composition** | Perfluorinated polyethylene backbone with pendant side chains terminating with sulfonic acid groups | Crosslinked sulfonated copolymers of vinyl compounds, cast in sheet form, homogeneous films on synthetic reinforced fabrics |
| **Membrane density (H-form)** | $d_{dry} = 2.02$<br>$d_{wet} = 1.64$ | $d_{dry} = 0.87$<br>$d_{wet} = 1.14$ |
| **Membrane thickness** | 0.21 mm (wet state) | 1.17 mm (wet state) |
| **Water content** | 33% dry resin<br>$\lambda = 22$ $H_2O/SO_3^-$ | 43% dry resin<br>$\lambda = 19$ $H_2O/SO_3^-$ |
| **Ion exchange capacity** | $0.91$ meq g$^{-1}$ (dry resin) | $2.44$ meq g$^{-1}$ (dry resin) |
| **Ion conductivity $\kappa$** | H-form: $1.2\times10^{-1}$ S cm$^{-1}$ (wet)<br>K-form: $3.0\times10^{-2}$ S cm$^{-1}$ (wet) | H-form: $2.0\times10^{-2}$ S cm$^{-1}$ (wet)<br>K-form: $4.2\times10^{-3}$ S cm$^{-1}$ (wet) |
| **Water transference coefficient $t_{H_2O}$** | H-form: 2.6 (wet)<br>K-form: 5.3 (wet) | H-form: 2.5 (wet)<br>K-form: 11.3 (wet) |
| **Applications** | Separation, chlor-alkali electrolysis, fuel cells, sensors, etc. | Separation, electro-dialysis, desalination, food processing, etc. |

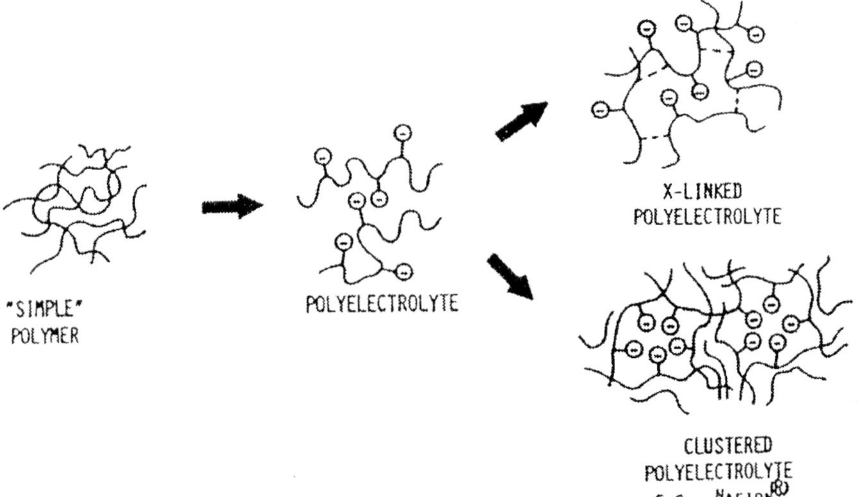


Figure 2. Crosslinked polyelectrolyte gels and phase-separated ionomers with anionic side chains (with permission from Plenum Press).

## 2.2. EXPERIMENTAL METHODS FOR IONOMER STRUCTURE STUDIES

Several methodologies for structure analyses of perfluorinated ionomers were presented by many authors. Wide-angle X-ray diffraction analysis (WAXD), small angle X-ray scattering (SAXS), small angle neutron scattering (SANS) and other diffraction methods made it possible to characterize crystalline structures and aggregation of ionic clusters in ionomers [10,12]. The size of ionic clusters was estimated from the data of diffraction peaks and observing scattering length due to the difference in the electron density between ion cluster regions and fluorocarbon matrices. Normally two scattering peaks are handled, the crystalline peak and the ionomer peak, the second often superimposed on a broad amorphous halo. The effects of equivalent weight, hydration and cation forms have been discussed to elucidate the structure model. However, the estimated morphology and the size of domains are largely model dependent, and there have been usually much debate concerning the exact morphology of Nafion® polymers [14].

Thermal analyses, especially differential scanning calorimetry (DSC) was used to elucidate the phase transition phenomena in the polymer, e.g., glass transition temperature, exo- and endo-thermal peaks of water melting in the polymer, etc. Thermal movement of polymer chains and thermo-relaxation were discussed from which structural features of induced electrostatic aggregates and the effect of water was investigated. The state of water in the ionomer membranes, e.g., bound water and free water and their quantitative relationships were discussed [17].

Infra-red spectroscopy (IR) gave information on the chemical bonding state of the polymer, and piles of data have been collected [18-20]. From oscillation peaks of OH groups, hydrogen bonding was discussed and different states of water were discriminated, e.g., non-hydrogen bonded water in the periphery of hydrophilic domains, hydrogen bonding between water molecules and that between ion and water. Also from the oscillations of $-SO_3^-$ groups, formation of ion pairs between sulfonic acid groups and counter-ions were observed during the membrane drying process.

## 2.3. MODEL OF EISENBERG

Eisenberg first studied the problem of microscopic phase separation in polymers, and proposed the model of ion clustering where the ionic multiplets are separated by nonionic material [13,21]. Starting from the association of ions (ion-pairs, triplets, quartets, etc.) this model gave the insight of the factors that limit the size of the multiplets, and subsequently the cluster size was determined as a function of various chain parameters. The energetic balance between the electrostatic interaction force and the elastic force working in polymer chains was taken into account, but the crystallinity was neglected in the model. The cluster morphology was energetically favorable below the critical temperature $T_c$ where the elastic and electrostatic forces balance each other. The model predicted the number of fundamental ion pairs within a stable cluster and the average intercluster distance. In spite of some oversimplified assumptions, reasonable agreements with experiments were achieved.

## 2.4. MODEL OF MAURITZ ET AL.

In the earlier model in 1980, Mauritz et al. proposed the cluster model [13], dealing with the molecular energetics of cluster formation (figure 3). The interacting dipoles within a cluster, the loss of configurational entropy associated with the formation of ion cluster and the energy of deforming polymer chains are considered. The hydration energies of dipoles formed between sulfonic acid groups and the counter-ions are balanced by the elastic deformation energy and expanding energy opposing the surface tension of hydrophilic│hydrophobic domain interface. The unfortunate point of this model was the lack of information about polymer structures at that time, and some theoretical assumptions (e.g., Gaussian coil approximation for chain deformation) and usage of empirical parameters made the model complex and inappropriate.

Later they proposed the water sorption isotherm model in 1985, in which they took into accountthe balance of elastic force of polymer network with the osmotic swelling force estimated from water activity in the micro-solution within ionic clusters [22]. Water activity was derived considering hydrated water to ions and free- and bound-water population. The osmotic pressure was balanced by a polymer matrix contractile pressure. The cluster radius extension ratio, the volume fraction of the ionic cluster phase, etc. were predicted.

## 2.5. GIERKE'S INVERSE MICELLE MODEL

In 1981, Gierke et al. proposed a new model that can explain the high current efficiency of membranes for chlor-alkali electrolysis [23]. In their model,

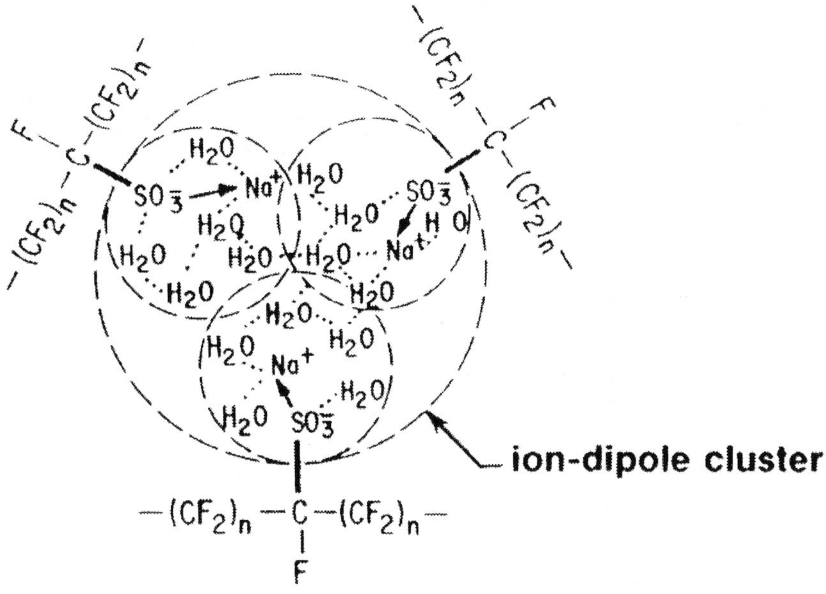



Figure 3. Cluster formed from the coalescence of hydrated ion-dipoles in PFSI (with permission from Plenum press).

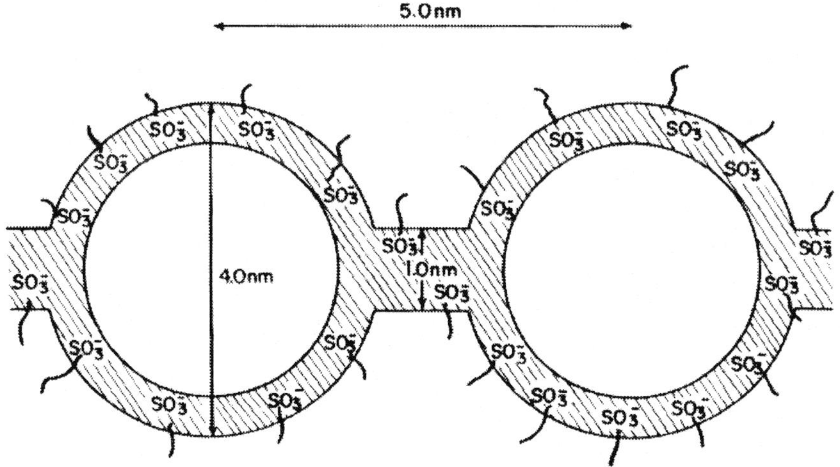



Figure 4. Gierke's cluster network model (with permission from the American Chemical Society).

spherical clusters of inverse micelle structures filled with water, sulfonated end groups and counter-cations are assumed that are interconnected with small cylindrical "pores", and forms the networks of hydrophilic regions (figure 4). From the SAXS studies of swelling membranes by varying the amounts of water, the size of clusters was estimated to be ca. 4 nm with channels of 1 nm size. This Gierke model, also called the "cluster network model", sufficiently explained the ion permselectivity and water transport in ionomer membranes, and most widely referenced in the literature. However, it should be noted that this model was not defined from observed structural properties per se, and many alternative morphologies have been proposed after that [14].

## 2.6. RODLIKE STRUCTURE MODEL

Recently a new structural model for the polymer aggregates is proposed, using SANS and SAXS experiments [24]. The aggregation of ionomer chains forms into elongated polymeric bundles with a diameter about 4 nm and length larger than 100 nm. This rodlike aggregates are surrounded by ionic groups and the water molecules. Upon dilution, the ionomer peaks in the scattering curves behave as rodlike particles and ribbonlike particles, at low and high polymer contents, respectively. The matrix peaks correlate with the crystallinity with a scale of 10 to 100 nm. On water swelling process, these aggregates follow two different regimes depending on the hydration: a lamellar dilution and a two-dimensional swelling.

# Chapter 3

# TRANSPORT MODELS OF IONOMERS

## 3.1. MODEL DESCRIPTION

Once the structure of PFSI is established by appropriate models, this could also be applied to explain the transport phenomena in microscopic levels. The elementary processes of transport occur in the phase-separated structures, which are microscopically interpreted as hydrophilic and hydrophobic domain structures. A molecular dynamic approach using computer simulations may be an ultimate solution to this system, but for application to macroscopic systems such as PEFC or other electrochemical systems, a simplified description from three-dimensional computation to one-dimensional description using proper transport models is inevitable. Both statistical and phenomenological theories can apply for such considerations, but the most convenient one is to modify the equations of liquid-like transport to one- or two-dimensional transport. Basic idea starts with the liquid-state media where ions and water molecules interact each other in the uniform three-dimensional atmosphere, and in the second step their interactions are confined in domain structures of polymers. In this sense a structure model of ionomers serves in two ways.

One is to adopt the liquid-like transport equations by using as "tortuous media", where the transport path is entangled and elongated due to the polymer network structure. The second uses the idea of "narrow channel structure", and the transport path is narrowed due to decreased cross sections for ion conduction in two-dimensional view. In both of these models, the transport behaviors are fit into simple equations with parameters such as "tortuosity" or "channel cross section", and it is discussed if the parameters have real physical meaning. The intermolecular forces, i.e., ion-ion, ion-water, ion-polymer and water-polymer interactions are derived if the model can explain the transport phenomena in simple manners.

## 3.2. STRUCTURE MODELS FOR ION TRANSPORT SELECTIVITY

Eisenberg et al. measured diffusion properties in Nafion® membranes, and found the activation energy of water diffusion was as high as that in pure water. The dynamic mechanical studies show that β relaxation (ionic regions) was much more affected by water than α relaxation (nonionic phase), indicating higher movement of molecules in the ionic clusters [25]. These results imply that transport properties are closely connected with the membrane polymer structures.

In 1981, Gierke et al. studied the microscopic structure of perfluorinated chlor-alkali electrolysis membranes with scattering and other experiments, and proposed "cluster network model" [23,26]. In this model hydrophilic ion cluster extends through the hydrophobic domain made of polymer main chains, so called the "inverse micelle structure". Using semi-phenomenological expression, the variation in cluster diameter with water content, equivalent weight, and cation form were discussed [27]. The percolative aspects of ion transport and the ion selectivity of membranes were described to explain the current efficiency during electrolysis.

Yeager et al. in 1981 assumed the third region besides the ion cluster region and the fluorocarbon backbone region, and called it "transitional interphase region" [28]. This region consists of polymer side chains, ions and water molecules, with relatively large voids where ions of larger sizes can be accommodated (figure 5). This model was called the "three-phase model", and was able to explain differences between sulfonic and carboxylic acid type ionomers where the intermediate region formed was much smaller in the latter than in the former. The third region appears as a transitional region between hydrophobic and hydrophilic regions, and works as the buffer for molecular movement.

Although the above two models are studied first to consider the ion selective functions of ionomer membranes, it should be mentioned that they extend to more practical transport models described below.

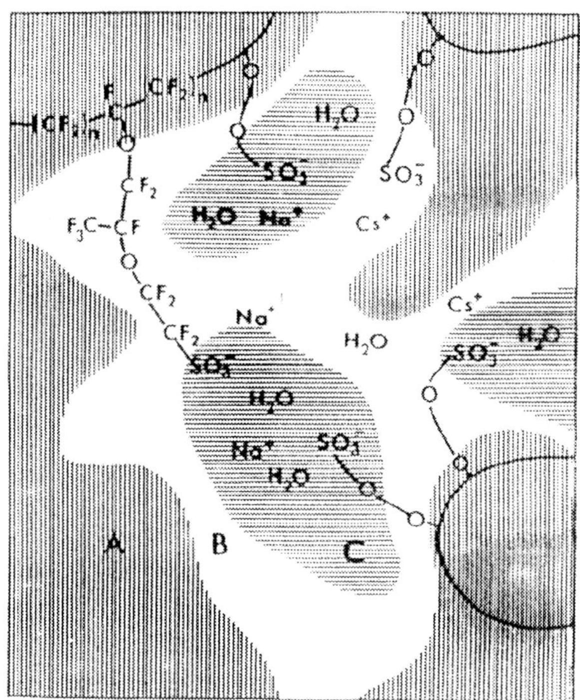


Figure 5. Three-region structural model of PFSI. A, fluorocarbon; B, interfacial zone; and C, ionic clusters (with permission from the Electrochemical Society).

## 3.3. FREE VOLUME MODEL

Ions can conduct only through the hydrophilic domains not the polymer crystalline regions, which are the barrier for conduction. This means that smaller cations can move more freely in the ion cluster because of smaller obstructions from tortuosity of the paths compared to larger cations. In 1955, Mackie and Mears proposed the theory in which they assumed a "pseudo-lattice" made of polymer regions and hydrophilic regions on the lattice points [29]. Compared with the diffusion in ionic solutions, diffusivity of ions is limited by the probability of formation of paths of hydrophilic domains connected on the lattice [30]:

$$D_i = D_i^0 f(V_p) \tag{1}$$

here $D_i$ and $D_i^0$ are the diffusion coefficients of ion i in the polymer and in liquid phases, respectively, and $V_p$ is the volume fraction of polymer crystalline region. In the lattice point model, $f(V_p)$ is given for $V_p < 0.5$ by

$$f(V_p) = \frac{(1-V_p)^2}{(1+V_p)^2} \tag{2}$$

Yasuda et al. in 1968 deduced $f(V_p)$ as follows, based on the free volume theory [31]:

$$f(V_p) = \exp\left(\frac{-bV_p}{1-V_p}\right) \tag{3}$$

In free volume theory, the ions can transfer when a new void is formed in the polymer chains to accommodate the ion during thermal motions.

Yeager et al. explored the dependence of self-diffusion coefficients of water and ions on membrane water content, using free volume approach [28]. Figure 6 shows the results for diffusion coefficients of water and ions in Nafion® 120 membranes. The experimental fact that activation energy of diffusion in the membrane does not differ much from that in the solution gives the rationale for this theory.

## 3.4. ION CHANNEL MODEL

In cluster network model, narrow channels (about 1 nm diameter) connect the ion cluster regions, and this "pore" makes the barrier for the ion conduction. The ions transfer in the hydrophilic domains in the ionomer, which is assumed as the liquid-like medium. In the Gierke's model, the relative size of diffusing species against the effective size of the pore may determine the diffusion coefficient. Also it is perceived that in order for the ion and water to move along the continuous path, there should be certain threshold volume fraction of ion clusters in the membrane. This can be handled by the concept of percolation [27].

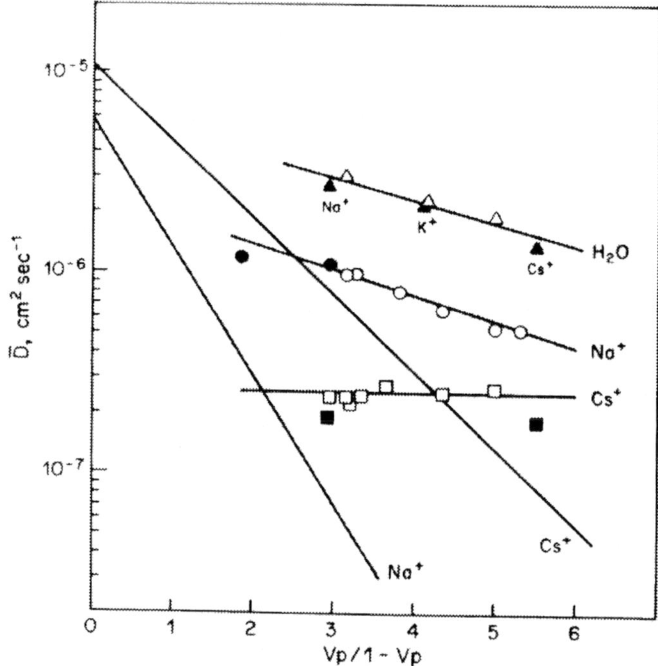


Figure 6. Logarithm of self-diffusion coefficient vs. polymer fraction function $V_p/(1- V_p)$ for Nafion$^®$ 120 at 25 °C. $Na^+$ and $Cs^+$ lines without data points are for poly(styrenesulfonate) (with permission from the Electrochemical Society).

In the percolation theory, the ion conductivity σ obeys the relation

$$\sigma = \sigma_0 \left( c - c_0 \right)^n \tag{4}$$

where $c$ is the volume fraction of water in the polymer, i.e., the ions start to conduct above the threshold value of the hydrophilic regions. In figure 7, the slope is obtained as $n = 1.5$, which is allegedly in agreement with three-dimensional percolation theory [27]. The fact $c_0 = 0.1$ means that the ion clusters are not uniformly distributed in the polymer but exist in aggregated forms. When the water content in the polymer is reduced, the ion conduction path is disconnected as a result of the polymer reorientation, and the ion-pair formation may start to occur.

## 3.5. OTHER MODELS

Verbrugge et al. developed a macrohomogeneous model of membrane transport starting from Nernst-Planck equation with convection terms [32]. They opposed the cluster-network morphology, and proposed an array of capillary pores with uniform cross section. Convection within the pores was described using pressure and electric potential gradients where local viscosity was assumed as a parameter. The proton diffusion coefficient was found to follow Bruggeman equation.

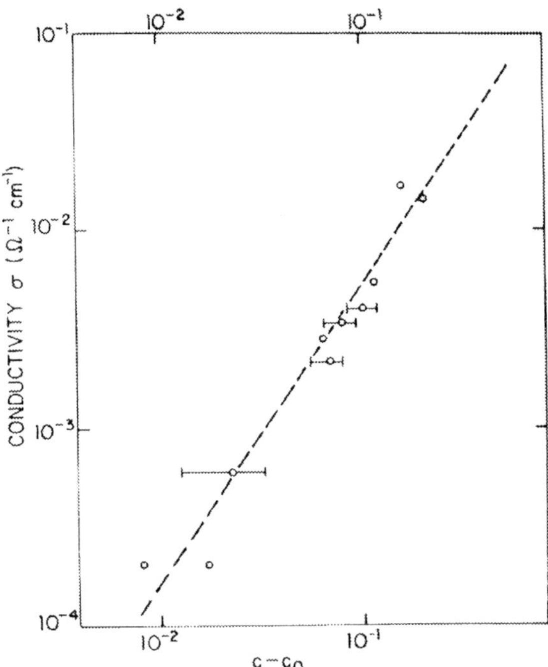

Figure 7. Log-log plot of conductivity vs. excess volume fraction of water in Nafion® membranes at room temperature (with permission from Elsevier Science).

$$D_H = D_H^0 \theta^B \qquad (5)$$

where $\theta$ is porosity and $B$ is Bruggeman exponent, which normally takes the value around 1.5. They suggested criteria for new membranes, i.e., it is desired that a membrane should have fixed charge density of about 1.2 mol dm$^{-3}$, pore diameter near 5.5 nm and a membrane porosity of 0.3.

Pintauro et al. established a cylindrical pore model for multicomponent ion and water transport in ion exchange membranes [33]. Sulfonic acid groups are assumed as fixed charge sites uniformly distributed on the pore wall surface. The flux equation of ion $N_i(x, r)$ inside the pore included gradient terms of electrochemical potential and volume flow.

$$N_i(x,r) = -u_i C_i(x,r)\nabla \mu_i(x,r) + C_i(x,r)v(x,r) \tag{6}$$

where $x$ and $r$ are axial and radial pore directions, $u_i$ and $C_i(x, r)$ are ion mobility and concentration, $\mu_i(x, r)$ the electrochemical potential and $v$ is volume flow. Ion/fixed charge interactions, water dipole orientation in the pore, ion hydration free energy changes were taken into account. By solving molar flux equation with local water dielectric constant in the pore, the flux components of diffusion, migration, convection and hydration were separated and the cation permselectivity during Donnan dialysis was predicted.

Paddison et al. proposed a statistical mechanical model of capillary transport of proton and water, with a nonuniform charge distribution on the pore walls [34]. The Hamiltonian of oxonium ion contains interactions with water molecules and the pendant side chains, which was solved with periodic boundary conditions in cylindrical geometry filled with water molecules. The proton diffusion coefficients at two levels of water content were predicted which was in agreement with the experimentally measured values.

*Chapter 4*

# Transport Analysis in Perfluorinated Ionomers

## 4.1. Basic Equations for Transport Analysis in the Membrane

The system of PEFC is modeled by the following scheme [35]:

$$H_2O(g_1), H_2(g_1), Pt \parallel HM, AM, H_2O \parallel Pt, H_2O(g_2), O_2(g_2) \tag{7}$$

where $\parallel \ \parallel$ indicates the membrane phase, HM and AM symbolize $H^+$ and $M^+$ ions combined with ion exchange groups in the polymer electrolyte membrane, Pt is either the anode or cathode catalyst to which gaseous species $H_2O$, $H_2$ and $O_2$ contact. The electrochemical cell (7) is assumed to be composed of two cells in series:

$$H_2O(g_1), H_2(g_1), Pt \parallel HM, AM, H_2O \parallel Pt, H_2O(g_2), H_2(g_2) \tag{8}$$

$$H_2O(g_2), H_2(g_2), Pt \parallel HM, AM, H_2O \parallel Pt, H_2O(g_2), O_2(g_2) \tag{9}$$

The cell (8) is considered as a concentration cell while the cell (9) is a chemical reaction cell. The cell (8) is an irreversible system with concentration gradients, and the following phenomenological equations are assumed to describe the transport phenomena [35].

$$j_{HM} = -L_{11}\Delta\mu_{HM} - L_{12}\Delta\mu_{AM} - L_{13}\Delta\mu_{H_2O} - L_{14}\Delta\varphi \qquad (10\text{-}1)$$

$$j_{AM} = -L_{21}\Delta\mu_{HM} - L_{22}\Delta\mu_{AM} - L_{23}\Delta\mu_{H_2O} - L_{24}\Delta\varphi \qquad (10\text{-}2)$$

$$j_{H_2O} = -L_{31}\Delta\mu_{HM} - L_{32}\Delta\mu_{AM} - L_{33}\Delta\mu_{H_2O} - L_{34}\Delta\varphi \qquad (10\text{-}3)$$

$$j = -L_{41}\Delta\mu_{HM} - L_{42}\Delta\mu_{AM} - L_{43}\Delta\mu_{H_2O} - L_{44}\Delta\varphi \qquad (10\text{-}4)$$

here $j_{HM}$, $j_{AM}$, $j_{H_2O}$ and $j$, designate fluxes of $H^+$ ion and $A^+$ ion, water and electricity, respectively, across the membrane, and $\Delta\mu_{HM}$, $\Delta\mu_{AM}$, $\Delta\mu_{H_2O}$ and $\Delta\varphi$ are forces (chemical potential differences) corresponding to these fluxes. $L_{ij}$ are phenomenological coefficients, and the Onsager reciprocal relations hold between these coefficients.

$$L_{ij} = L_{ji} \qquad (11)$$

The relation $D_i = L_{ii}RT/C_i$ holds for diffusion coefficient of species i, and $\kappa = L_{44}F^2$ for electric conductivity. Under no concentration gradients through the membrane, it is written using Onsager relations as [35]

$$\left(\frac{j_i}{j}\right)_{\Delta\mu_i=0} = \frac{L_{i4}}{L_{44}} = t_i \qquad (12)$$

$t_i$ is called as the "transference coefficient" of species iM, and related to the "ion transference numbers" $t_{i^{n+}}$ for ions. For example, when the ion i has $n$ valence, the relation holds for the transference number of $i^{n+}$ as $t_{i^{n+}} = nt_{iM}$ [35].

## 4.2. ELECTROCHEMICAL METHODS FOR TRANSPORT ANALYSIS

### 4.2.1. Membrane Contact Potential

In order to elucidate the transport mechanism in the membrane, it is essential to obtain the transport parameters of ion and water with the help of electrochemical measurements. The important parameters are the ionic transference numbers for ionic species and water transference coefficient for water flux in electro-osmotic phenomena.

The electrochemical cell is set so that the total concentration of electrolytes is maintained equal at both sides of the membrane. Here two cation species $H^+$ and $Na^+$ exist as mixed ionic solutions (HCl and NaCl).

$$Ag \mid AgCl \mid H_2O, HCl, NaCl \parallel HM, AM, H_2O \parallel H_2O, HCl, NaCl \mid AgCl \mid Ag \qquad (13)$$

The difference in the electromotive force (*emf*) across the membrane, $\Delta\varphi$, is measured with a pair of $Ag \mid AgCl$ electrodes. When no current flows through the membrane, $\Delta\varphi_{j=0}$ is obtained as follows, using Eq. (10-4) together with Eqs. (11) and (12).

$$\Delta\varphi_{j=0} = -t_{H^+}\Delta\mu_{HM} - t_{Na^+}\Delta\mu_{NaM} - t_{H_2O}\Delta\mu_{H_2O} \qquad (14)$$

At the equilibrium condition between the membrane and the contacting solutions, the relation $\Delta\mu_{HM} = \Delta\mu_{HCl}$ holds across the membrane. Also it is assumed that there is very little concentration difference of water through the membrane ($\Delta\mu_{H_2O} = 0$). The Gibbs-Duhem equation:

$$x_{HCl}\Delta\mu_{HCl} + x_{NaCl}\Delta\mu_{NaCl} = 0 \qquad (15)$$

together with the relation for the ionic transference numbers, $t_{H^+} + t_{Na^+} = 1$, gives

$$t_{H^+} = x_{HCl}\left(\frac{\Delta\phi_{j=0}}{\Delta\mu_{NaCl}} + 1\right) = x_{HCl} - (1 - x_{HCl})\left(\frac{\Delta\phi_{j=0}}{\Delta\mu_{HCl}}\right) \qquad (16)$$

By use of the concentration cell (13) where HCl + NaCl solution of different composition is contacting at each side of the membrane, the ionic transference number of H$^+$ in the membrane, $t_{H^+}$, is obtained from Eq. (16) by measuring the electromotive force [36]. The present method has an advantage over Hittorf method in terms of precision and the ease of measuring and time [36].

A pair of Nafion® 117 membranes, each equilibrated with a solution of fixed ionic composition, are overlapped at one end, and the other ends are contacting the equilibrating solution with Ag/AgCl electrodes (figure 8) [37]. One of the membranes is a reference membrane, and the other is a test membrane contacting solutions of various kinds of $X_{HCl}$ levels. *Emf* measurements are performed after steady state potential is attained.

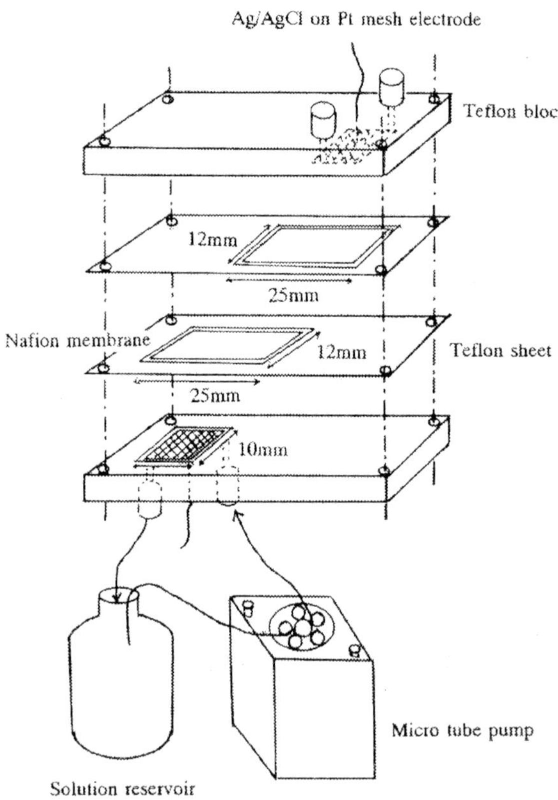


Figure 8. Schematic representation of the concentration cell used for the measurement of the ionic transference numbers in the membrane (with permission from Elsevier Science).

## 4.2.2. Streaming Potential Method

The electrochemical cell is set so that two identical mixed ionic solutions (HCl and NaCl) are contacting the membrane at both sides.

$$Ag \mid AgCl \mid H_2O, HCl, NaCl \parallel HM,AM,H_2O \parallel H_2O, HCl, NaCl \mid AgCl \mid Ag \qquad (17)$$

A differential pressure, $\Delta p$, is applied across the membrane and the resulting electromotive force is measured (figure 9). The chemical potential difference across the membrane is $\Delta \mu_i = V_i \Delta p$, where $V_i$ is the molar volume of species i, so that it is obtained:

$$\Delta \phi_{j=0} = -t_{H^+} \Delta p V_{HCl} - t_{Na^+} \Delta p V_{NaCl} - t_{H_2O} \Delta p V_{H_2O} \qquad (18)$$

The water transference coefficient, $t_{H_2O}$ can be obtained from the following equation

$$t_{H_2O} = -\left\{ \left( \frac{\Delta \phi_{obs}}{\Delta p} \right)_{j=0, \Delta c_i=0} + t_{H^+} V_{HCl} + t_{Na^+} V_{NaCl} + V_{el} \right\} / V_{H_2O} \qquad (19)$$

Here

$$\Delta \varphi_{obs} = \Delta \varphi_{j=0} - \Delta V_{el} \Delta p, \qquad \Delta V_{el} = V_{Ag} - V_{AgCl} \qquad (20)$$

and $V_{Ag}$ and $V_{AgCl}$ are molar volumes of Ag and AgCl, respectively. According to the irreversible thermodynamic theory, the electro-osmotic flow and streaming potential are equated with each other [35]:

$$-\left( \frac{j_V}{j} \right)_{\Delta p=0, \Delta c=0} = \left( \frac{\Delta \phi_{obs}}{\Delta p} \right)_{j=0, \Delta c=0} \qquad (21)$$

Measuring potential difference when the pressure is applied is much easier than measuring the volume flow accompanying the electric current. Implementing the streaming potential experiment has the advantage over the electro-osmosis experiment in terms of the ease and accuracy of $t_{H_2O}$.

In an electrochemical cell for streaming potential measurements, it is often encountered that after the pressure pulse is applied, the transient curve increase with time $t$, not the step rise of *emf*, occurs as shown in figure 10. This is due to the "concentration polarization" at the membrane|solution interface. This is expressed as follows [38]:

$$Emf = a - b\sqrt{t} \tag{22}$$

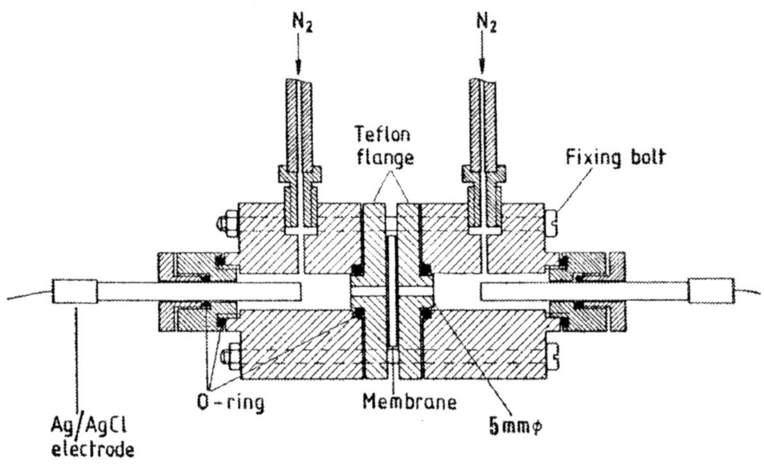


Figure 9. A scheme of the streaming potential measuring apparatus (courtesy of Prof. Signe Kjelstrup, Department of Chemistry, Norwegian University of Science and Technology).

The intercept of *emf* at time $t \to 0$, is defined as $E$, and the value of $\Delta\varphi_{obs} = EF$ is put in Eq. (19) to obtain $t_{H_2O}$. Also from the slope $b$ of *emf* vs. time curve, the water permeability $L_p$ is calculated using the equation:

$$L_p = -bF / (f\Delta p)$$

$$f \equiv \frac{4RT}{\sqrt{\pi}} \left\{ \left[ t_H - t_H(l) \left( D_{HCl}^{-1/2} - \frac{1}{n} D_{ACl_n}^{-1/2} \right) + t_{Cl}(l) \left[ \frac{1}{n} D_{ACl_n}^{-1/2} + \frac{c_{HCl}^0 D_{HCl}^{-1/2} + nc_{ACl_n}^0 D_{ACl_n}^{-1/2}}{c_{HCl}^0 + nc_{ACl_n}^0} \right] \right] \right\} \tag{23}$$

where $t_{H^+}(l)$ is transference number of $H^+$ in the solution, $D_{HCl}$, $D_{ACl_n}$, $c^0_{HCl}$ and $c^0_{ACl_n}$ are diffusion coefficients and concentrations of HCl and $ACl_n$ in the solution, respectively.

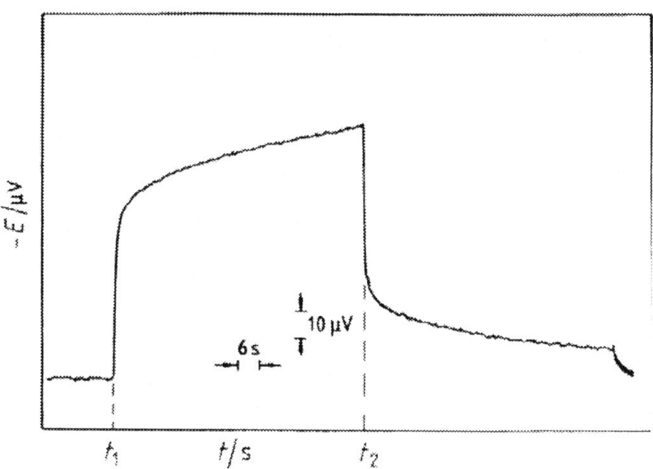


Figure 10. *Emf* as a function of time in the cation exchange membrane CR61 AZL 386 in 0.03 mol dm$^{-3}$ KCl at $\Delta p$ = 200 mmHg. At time $t_1$ the pressure is applied and at time $t_2$ removed (with permission from Elsevier Science).

A good point of this method is that two kinds of parameters, water transference coefficient $t_{H_2O}$ and water permeability $L_p$, (m$^4$ J$^{-1}$ s$^{-1}$) can be obtained in a single measurement. $t_{H_2O}$ arises due to the water movement associated with the charge transport, while $L_p$ arises due to the pressure-driven flow of water in the membrane [38]. The experiments are performed for Nafion 115 in deaerated solutions at 25 °C. Pressure is applied by N$_2$ gas with pressure control by model 250C-1-D of MKS Instruments, Inc., USA [37].

## 4.2.3. Impedance Measurement for Membrane Ionic Conductivity

A special electrochemical cell is employed for the ionic conductivity measurements, because the perfluorinated membranes have very high conductivity and the resistance for perpendicular direction of thin membranes shows often negligibly small values. Figure 11 depicts the Teflon cell where a pair of 10 mm wide platinum foil electrodes deposited with Pt black contact the membrane from both sides with separation of 5 mm. The membrane impedance is measured for the lateral direction of the membrane and the specific conductivity of the membrane $\kappa$ is obtained from the real part of the impedance $R$ ($\Omega$).

$$\kappa = \frac{s}{Rwl} \tag{24}$$

where $w$ is the width of the electrode (cm), $s$ is the separation of a pair of electrodes contacting the membrane (cm) and $l$ is the thickness of the membrane (cm).

A Solartron S-1260 frequency response analyzer (Solartron Instruments, UK) is used for measuring the impedance of the membrane in the frequency range 1 kHz to 0.1 Hz, with 20 mV ac modulation over a frequency range from $10^7$ to $10^3$ Hz at the open-circuit potential [39].

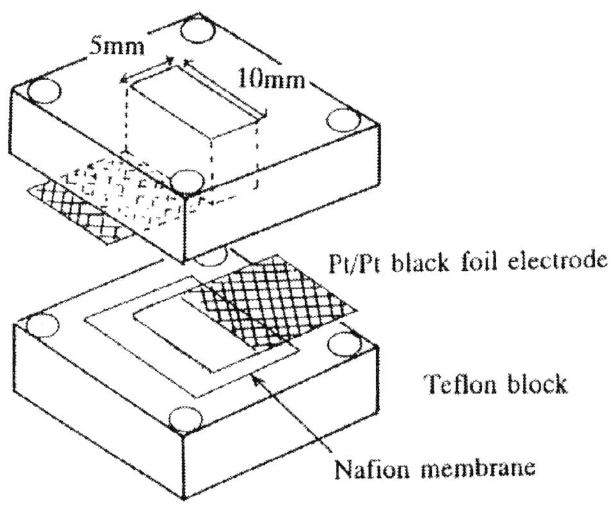


Figure 11. The Teflon® cell for membrane impedance measurements (with permission from the Electrochemical Society of Japan).

## 4.2.4. Ionic Mobility

The ionic transference number of $H^+$ in the membrane is given as [40]

$$t_{H^+} = \frac{x_{HM}\,u_{H^+}}{x_{HM}\,u_{H^+} + x_{AM}\,u_{A^+}} \tag{25}$$

Also the ionic conductivity of the membrane is expressed as follows, using the concentration of cation exchange sites $c_{SO_3^-}$ and the mobility of ion i, $u_i$:

$$\kappa = Fc_{SO_3^-}\left(x_{HM}\,u_{H^+} + x_{AM}\,u_{A^+}\right) \tag{26}$$

from which the mobilities of ionic species in the membrane, $u_{H^+}$ and $u_{A^+}$, are calculated [41].

$$u_{H^+} = \frac{\kappa t_{H^+}}{Fc_{SO_3^-}\,x_{HM}} \tag{27-1}$$

$$u_{A^+} = \frac{\kappa(1-t_{H^+})}{Fc_{SO_3^-}\,(1-x_{HM})} \tag{27-2}$$

The concentration of sulfonic acid group in the membrane, $c_{SO_3^-}$, is expressed as follows:

$$c_{SO_3^-} = \frac{d_{dry}}{\left(EW'\chi_V\right)} \tag{28}$$

where $EW'$ is the equivalent weight value corrected for the exchanged cations $A^+$ in place of $H^+$, $M_A$ being the atomic weight of the element A:

$$EW' \equiv EW + \left(M_A - 1\right)\left(1 - x_{HM}\right) \tag{29}$$

and $\chi_V$ is the ratio of membrane volume between wet and dry states: $\chi_V = V_{wet}/V_{dry}$. The equivalent weight $EW$ is a measure of the density of ion exchange groups, and is defined as the membrane dry mass (g) that contains one mol of the ion exchange groups (inverse of the ion exchange capacity *IEC* (meq g$^{-1}$), which can be measured by titration method):

$$EW=1000/IEC \qquad (30)$$

## 4.3. CATION EXCHANGE EQUILIBRIUM IN PERFLUORINATED IONOMERS

Because the composition of cationic species inside the membrane in binary cation systems is often different from that in the environment, at first the partition isotherm should be established. Ionic composition in the membrane is determined by X-ray fluorescence spectroscopy in dry state. A Seiko Electric Co. model SEA2010 X-ray fluorescence spectroscopic analyzer is used for component analysis, and elements A (A = Ca, Fe, Ni, Cu), S and Cl are picked from the spectra [41,42]. The cationic site fraction of H$^+$ in the membrane, $X_{HM}$, is calculated from the atomic ratio [A]/[S] in the spectra.

$$X_{HM} = 1 - \frac{[A]/[S]}{([A]/[S])_{X_{HCl}=0}} \qquad (31)$$

where [A] and [S] are intensities of the elements A and S in the spectra, and the value in the denominator means the ratio [A]/[S] at $X_{HCl} = 0$. EPMA (electron probe microanalysis) is also employed for the Na$^+$ content determination using Eq. (31) [37]. A JEOL electron prove micro-analyzer model JXA-8800M is used to obtain the energy dispersion spectra. The analysis is further verified by inductively coupled plasma emission spectrography (ICP, Thermo Jarreel Ash, model IRIS/AP) after the ion was extracted into the liquid solution by soaking the membrane in 0.1 mol dm$^{-3}$ HCl [42].

The equilibrium constant $K_{ex}$ of the exchange reaction between the solution phase and the membrane phase, HCl(aq) + AM = ACl(aq) + HM, where M denotes the cation exchange site, is defined as follows [43]:

$$K_{ex} = \frac{x_{ACl} x_{HM}}{x_{HCl} x_{AM}} \qquad (32)$$

The interaction between neighboring cations on the cation exchange sites can be discussed based on the cation mixing process in the membrane. For a system of monovalent-monovalent cations in the membrane, HM and AM, thermodynamic equilibrium constant $K_{th}$ is defined for the reaction HCl(aq) + AM = ACl(aq) + HM as follows:

$$K_{th} = \frac{a_{ACl}a_{HM}}{a_{HCl}a_{AM}} = K' \frac{\gamma_{HM}}{\gamma_{AM}} \qquad (33.1)$$

$$K' \equiv \frac{a_{ACl}x_{HM}}{a_{HCl}x_{AM}} \qquad (33.2)$$

where $a_{HCl}$ and $a_{ACl}$ are activities of HCl and ACl in the solution, respectively, and $\gamma_{HM}$ and $\gamma_{AM}$ are activity coefficients of HM and AM in the membrane, respectively.

## 4.4. STATE OF WATER IN THE MEMBRANE

The membrane performance is very sensitive to the state of water inside the membrane, due to the specific structure of perfluorinated ionomer membranes. The cation conductivities in the membrane are closely related to hydrated water around the cation and sulfonic acid groups, local structure of hydrophilic domains and the amount of bulk-like water filling the ionic channels. The state of water, therefore, is a very important parameter in order to discuss the mechanism of ion transport in the membrane, and should be studied from various viewpoints.

Water content in the membrane is determined by gravimetric method. A membrane sample is taken from the equilibrating solution, blotted lightly with tissue paper, and weighed immediately ($W_{wet}$). The sample is then dried completely in vacuum at 110 °C for 12 h [37,39]. After the sample is cooled in a desiccator, it is weighed ($W_{dry}$), and from the weight difference the water content was calculated as the number of water molecules per cationic site $\lambda = n_{H_2O} / n_{SO_3^-}$:

$$\lambda = \frac{(W_{wet} - W_{dry})EW'}{18W_{dry}} \tag{34}$$

The volume fraction of water $\theta$ in the membrane is determined from the volume of the membrane in dry and wet states ($V_{dry}$ and $V_{wet}$, respectively, which are measured by a micrometer):

$$\theta = \frac{V_{wet} - V_{dry}}{V_{wet}} \tag{35}$$

The density of the membrane in wet state, $d_{wet}$ is also calculated by a gravimetric method: i.e., the weight of the wet membrane, $W_{wet}$ is divided by the volume, $V_{wet}$:

$$d_{wet} = W_{wet}/V_{wet} \tag{36}$$

The phase transition of water absorbed in the membrane is measured by differential scanning calorimeter (Seiko Co., Ltd. Model DSC 220C). Membrane pieces are put in an aluminum pan, and DSC curves are obtained by cooling the sample from room temperature to $-120$ °C at the scanning rate of 1 °C min$^{-1}$. The amount of freezing water in the membrane is calculated from the peak area of freezing, using the enthalpy of possible polymorphism of ice. The amount of non-freezing water is obtained as the difference between the total water content in the membrane and the amount of freezing water [39].

## 4.5. INFRA-RED SPECTROSCOPY

The interaction between cations, cation and sulfonic acid groups and water in the membrane phase is discussed by measuring chemical bonding states through IR spectroscopic analyses. Very thin recast membranes of thickness ca. 20 µm are used to perform Fourier transform infra-red (FT-IR) spectroscopy experiments [39,44]. Membranes are exchanged with various alkali and alkaline earth metal cations by equilibrating in solutions of various cationic chloride salts, and their chemical bonding states are measured using FT-IR spectrometer (Bio-Rad Lab., Inc. model FTS-60A) in the wavenumber range 4000-700 cm$^{-1}$ with a resolution of 1-4 cm$^{-1}$.

## 4.6. DIFFUSION MEASUREMENTS BY NMR SPECTROSCOPY

NMR spectroscopy is one of the most important methods for studying molecular dynamics. The NMR parameters including the spin-lattice relaxation time $(T_1)$, the spin-spin relaxation time $(T_2)$, line width and spectral pattern are sensitive to molecular motions and can be used to clarify reorientational and translational motions under inter- and intra-molecular interactions. Among the molecular motions, translational diffusion is closely related to the functions of materials. The timescale of NMR measurements of translational diffusion is determined by the length of the pulse sequence and is typically in the range of 20 to 200 ms which spans the timescale needed to study many interesting chemical phenomena.

The direct measurement of translational motions by applying a static magnetic field gradient (FG) during a spin-echo experiment was proposed in the 1950s. The experiments faced difficulties, however, due to the inherent limitations of conducting the entire NMR measurement (including signal acquisition) in the presence of a static field gradient with relatively low field strengths provided by iron-magnets. Recent advances in diffusion measurements were stimulated by the development of the pulsed field gradient (PFG) variant of the static gradient method; pulsed-field-gradient spin-echo (PGSE) NMR. The technology to generate short sharp magnetic gradient pulses was also a key factor in the development of high resolution NMR. Here we will focus on translational (or self-) diffusion measurements using the PGSE-NMR method to apply for the molecular and ion diffusion in membranes, solution and polymer electrolytes.

The simplest PGSE sequence is shown in figure 12 where a Hahn echo sequence is modified to include two PFG pulses into each $\tau$ delay intervals. Basically the Hahn echo sequence has been important to measure $T_2$ and the interval time $\tau$ closely related to $T_2$ is fixed for the PGSE-NMR. Before the signal decays by the $T_2$ process, the attenuation of the echo signal induced by the PFG is measured. Here $g$ is the height of the PFG (typically 0.01 to 20 T m$^{-1}$, 10 to 2000 gauss cm$^{-1}$), the $\delta$ is the duration of the PFG pulse (typically 0.1 to 10 ms) and $\Delta$, which defines the timescale of the diffusion measurement, is duration between the leading edges of the gradient pulses. The maximum usable $\Delta$ (typically 20 ms to 1 s) is determined by $\tau$ and ultimately the $T_2$ of the species being measured. The attenuation $(E)$ of the amplitude of the spin-echo signal $(S)$ can be related to the experimental parameters. The translational self-diffusion coefficient $(D)$ is calculated using the Stejskal equation [45],

$$\ln(E) = \ln(S / S_{g=0}) = -\gamma^2 g^2 D\delta^2 (\Delta - \delta / 3) \qquad (37)$$

A diffusion measurement is performed by measuring $E$ as a function of $\delta$ or $g$ and regressing Eq. (37) onto the attenuation data, where $\tau$ is kept constant to allow attenuation due to relaxation to be normalized out. The effective PFG strength is determined by $g \times \delta$. At present, two types of experiments are possible to control the PFG strength; changing $\delta$ under the fixed PFG or changing $g$ under the fixed $\delta$. If the setting parameters are proper, the both experiments give the same $D$. When the diffusion plot following to Eq. (37) does not fit linear relations, the diffusion phenomena are complicated like restricted diffusions with many varieties or multiple diffusions consisting of multi-components. The physical parameter $D$ can be obtained only from the linear Stejskal plot.

## Hahn echo sequence

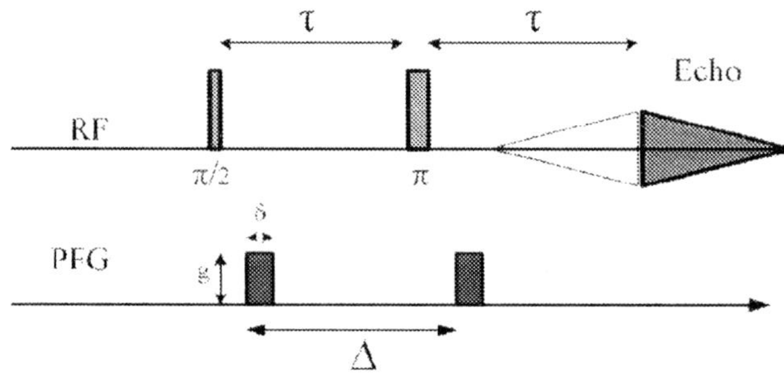


Figure 12. The modified Hahn spin-echo sequence inserted two equivalent gradient pulses and the attenuation of the echo signals is observed.

Another commonly used pulse sequence is the stimulated spin-echo (STE) pulse sequence (figure 13), which has particular advantages for systems where the condition $T_2 < T_1$ holds. The relaxation process in the interval $\tau$ is governed by the $T_2$ process. The spins relax with $T_1$ process during the $\tau_1$. Since the phase cycling of the measurement is 16 in the STE pulse sequence, the accumulation must be multiples of 16. Furthermore, the sensitivity is a half of that of the Hahn echo sequence, the STE pulse sequence is not preferable. However, since the diffusion of the whole molecule is independent of the $T_1$ and $T_2$ of the magnetic resonance, the self-diffusion coefficients should be exactly the same measured by either the Hahn echo sequence or the STE sequences if the measuring conditions are proper. The $^1$H NMR spectrum of organic molecules shows often many peaks due to the chemical shifts with different $T_1$ and $T_2$. Since the $D$ of a compound under study is the parameter for the displacement of the gravity center, all signals should have the same $D$ value, except for the existence of specific interactions like hydrogen-bonding [46].

Usually the calibration of the height of the PFG is necessary before measuring the $D$ values. For the smaller $g$, $H_2O$ is available for which $D_{H_2O}$ at 25 °C $= 2.3 \times 10^{-9}$ m$^2$ s$^{-1}$ [47] and $^2$H NMR of $D_2O$ is convenient for the larger $g$ due to the smaller gyromagnetic ratio $\gamma$ of $^2$H resonance (about 1/6 of the $^1$H resonance). The range of $D$ values accessible depend on the PFG height of the probe and for enough $g$ (20 Tm$^{-1}$ maximum) the ranges are from $10^{-8}$ to $10^{-13}$ m$^2$ s$^{-1}$.

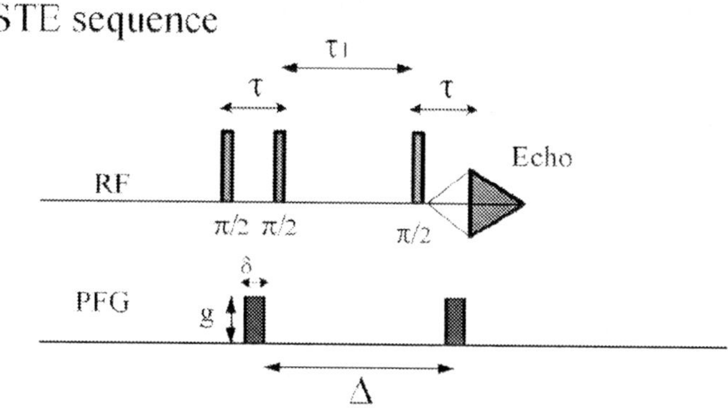


Figure 13. The stimulated echo (STE) pulse sequence to measure the diffusion which is useful for systems having $T_2 < T_1$.

The setting parameter $\Delta$ is important. When a sample under study is homogenous, the diffusion follows the Frick' low and the $D$ becomes independent of the $\Delta$. By using smaller $g$, the echo decays can be observed with the larger setting of the $\Delta$, but the signal sensitivities become smaller due to the $T_2$ and/or $T_1$ effects. The convection effects are enhanced by the longer $\Delta$ with the variable temperature measurements. In order to check the validity of the $D$ values, it is important to measure the $D$ values by varying the $\Delta$ values [48].

It is important to indicate that a self-diffusion coefficient measured by NMR is a physical constant. The concept of a self-diffusion coefficient $D$ is fundamental in physical chemistry. $D$ is often introduced using the Stokes-Einstein equation,

$$D = \frac{kT}{6\pi\eta r_s} \tag{38}$$

where $r_s$ is the Stokes (effective hydrodynamic) radius of the diffusing particle, $\eta$ is the solvent viscosity, $k$ is the Boltzmann constant. Implicitly, in the derivation of Eq. (38) the assumptions are that the diffusing particle sees the solvent as a continuum (i.e., the solvent molecules should be much smaller than the diffusing species) and that the diffusing species do not interact (i.e., infinite dilution). In the early stage of NMR diffusion study, the NMR diffusion coefficients were confirmed to be consistent with the data measured by using radio-active materials, and nowadays the diffusion phenomena are studied dominantly by NMR. Electrochemically, the relationships between the ion diffusion and ionic conductivity are interesting.

One of the most important theories in the electrochemistry is the Debye-Hückel theory which has been generally verified experimentally near infinite dilution in solution electrolytes. The equivalent ionic conductivity $\Lambda_D$ at infinite dilution is given by the Nernst-Einstein equation

$$\Lambda_D = \frac{Ne^2}{kT}(D_+ + D_-) \tag{39}$$

where $N$ is the number of cations per cm$^3$, $e$ is the electronic charge and $D_+$ and $D_-$ are the diffusion coefficients of the cation and anion, respectively. The ionic conductivities are measured by the electrochemical AC method and $D_{Li}$, $D_{anion}$ and $D_{solvent}$ by the PGSE-NMR method near the infinite dilution in organic electrolytes containing lithium salts, and extrapolated to infinite dilution [49]. The extrapolated values of $\Lambda_{AC}$ and $\Lambda_D$ agree well in the four solution electrolyte systems with two solvents and two lithium salts. In the practical solution electrolytes on the other hand, the calculated $\Lambda_D$ from NMR is always larger than $\Lambda_{AC}$, because the NMR can not distinguish the charged and paired ions but the $\Lambda_{AC}$ are obtained from the charged ions. Then the relative ratio $\Lambda_{AC}/\Lambda_D$ can give the degree of the ion dissociation. Although the ionic conductivity for the charged ions can not distinguish individual ions like cation or anion, multinuclear PGSE-NMR method can afford the ion diffusion independently without the knowledge of ion pairing. Generally, the ionic conductivity and the ion diffusion are well correlated [50].

It is known that an averaged distance in the three dimensions is given as follows

$$\overline{\Delta r^2} = 6D\Delta \tag{40}$$

The variable range of the $\Delta$ is usually limited 20 to 200 ms due to the $T_2$ and the reliable $D$ values for the homogenous systems measured by PGSE-NMR are $10^{-8}$ to $10^{-13}$ m$^2$ s$^{-1}$. The averaged distance which NMR can afford information is μm ($10^{-6}$ m) order in a space. The ionic conductivity is vague for the time scale, and also the Stokes-Einstein equation does not include the concept of time of diffusion explicitly. The diffusion coefficients obtained by the PGSE-NMR include the time scale and the space distances.

## 4.7. METHANOL PERMEABILITY

Methanol permeability is measured using a permeation cell composed of a pair of glass compartments, each with a capacity of approximately 80 ml [51]. The membrane sample is sandwiched between two glass compartments. The membranes are equilibrated in deionized water for at least 24 h prior to measurements. One compartment ($V_1$) is filled with 10 wt% methanol-water solutions. The other one ($V_2$) is filled with pure deionized water. Methanol diffuses from compartment $V_1$ to $V_2$ by the concentration difference between the two compartments, which are well-stirred during the test. The concentration of methanol in the receptor compartment, $C_2(t)$ is measured continuously using IR absorption methanol sensor (FCD-100, FC Development Co. Ltd., Hitachi) through a peristaltic pump.

Methanol permeability $P_M$ (cm$^2$ s$^{-1}$) is calculated from the slope of the $C_2(t)$ vs. time curve in the receptor compartment.

$$C_2(t) = \frac{A}{V_2} \frac{DK}{L} C_1(t_0)(t - t_0) \tag{41.1}$$

$$P_M \equiv DK = \frac{1}{A} \frac{C_2(t)}{C_1(t_0)} \frac{V_2 L}{t - t_0} \tag{41.2}$$

where $C_1(t_0)$ and $C_2(t)$ are the two methanol concentrations in compartment 1 and 2 at the initial and time $t$, respectively, $A$ and $L$ are the membrane area and thickness. $D$ and $K$ are the methanol diffusivity and partition coefficient between the membrane and the adjacent solution phases, respectively.

# Chapter 5

# ION AND WATER TRANSPORT CHARACTERISTICS IN PERFLUORINATED IONOMERS: MOLECULAR INTERACTIONS

## 5.1. DISCRETE CHANNEL STRUCTURE OF PERFLUORINATED IONOMERS

The transmission electron micrographs (TEM) of Nafion® membranes in figure 14 give specific sulfur imaging, where nonrandom distribution of sulfonic acid groups is shown as ~5 nm clusters in the polymer [52]. Similar morphology was observed for $EW$ 1200 Nafion® membranes substituted by $Ag^+$ and $Sn^{2+}$ ions, with metallic clusters of diameters 3-10 nm [23]. These TEM images strongly suggest that the ion conduction path exists in a manner of non-randomly distributed networks in the polymer and confined in a group of discrete domains. This morphology is reminiscent of the cluster network model discussed in section 2.5, and indeed the size of clusters fits the model.

In figure 15, specific resistances of Nafion® membranes exchanged with several kinds of alkyl-ammonium ions are plotted as functions of the membrane water content or the cation diameters [53]. When the cation size becomes as large as 1 nm, membrane resistance steeply increases, indicating that the channel pore size limits the movement of large cations (so called "plugging effect"). This observation is in good accordance with Gierke's cluster network model, where the ionic cluster of size 4 nm is interconnected by channels of diameter 1 nm. It is also noted that the amount of water decreases rapidly as the size of alkyl-ammonium cations increases (becomes more hydrophobic), and this causes high membrane resistance.

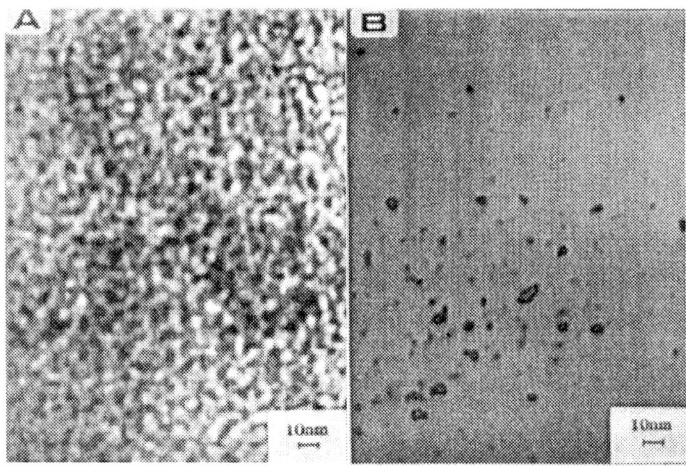

Figure 14. (A) Zero-loss image of a H-form Nafion® film. (B) Same field as (A), showing specific electron energy loss spectroscopy imaging for sulfur (with permission from the American Chemical Society).

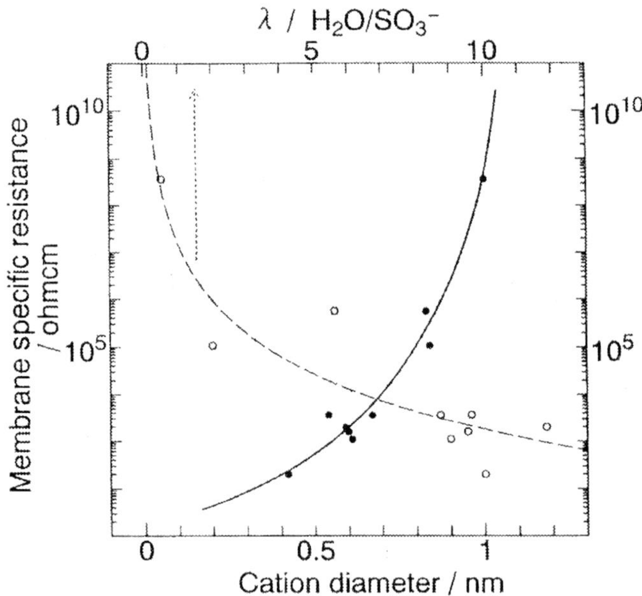

Figure 15. Membrane specific resistance of Nafion® 117 plotted as a function of the diameter (or length) of alkyl-ammonium cations or membrane water content (with permission from the Royal Society of Chemistry).

## 5.2. FROM LIQUID TO AGGREGATED STATES OF IONOMERS

The formation process of the discrete ionic channels during the polymer network formation of perfluorinated ionomer is an interesting theme to be investigated. We started to prepare the polymer suspended solutions with varying the concentration and see the evolution of gelling processes. At the low concentration of the ionomer solution, loose polymers would interact molecularly and then make aggregations of micelle-like structures. This initial gelling process was investigated by liquid rheology and electric conductivity measurements.

Nafion® membranes were dissolved in ethanol/water (50/50) solvent at 230 °C for 4 h in a pressure vessel. After vacuum evaporation Nafion powder was obtained, which was then dissolved in ethanol or other solvents, filtered and concentrated to 10 mg mL$^{-1}$. The equilibrium aggregation in the liquid state was studied by dynamic light scattering [54], and it was found that the main peak in the distribution function located in $10^3$ nm, and a smaller peak at 50 nm. These two types of aggregates in Nafion solutions may be categorized through thermodynamic equilibrium: one is large aggregate with hydrodynamic radius of $10^3$ nm and the second with the size 50 nm of loose polymer, dissolved molecularly.

Rheological and electric impedance measurements were performed through the process of gelling from diluted to concentrated Nafion solutions [55]. Interestingly, in sodium form Nafion, percolation behavior was observed in the electric conductivity as the Nafion volume fraction in the solution increased to 2.5 volume %, while in proton form Nafion no electrical percolation occurred. This percolation behavior is explained with the "fringed rod model" that is assumed for the transition between the micellar structure in solution and reverse micellar structure in the swollen membranes [56]. During the gelling, sulfonic acid groups are distributed over the whole network in the rod-like shape aggregates so that the charge carrier sites are connected above the threshold (figure 16). In H-form Nafion, this connection of the ionic conduction path becomes more difficult because of the larger specific volume or loose network structure, presumably due to high water content, as indicated by the thermal analysis in section 5.5.

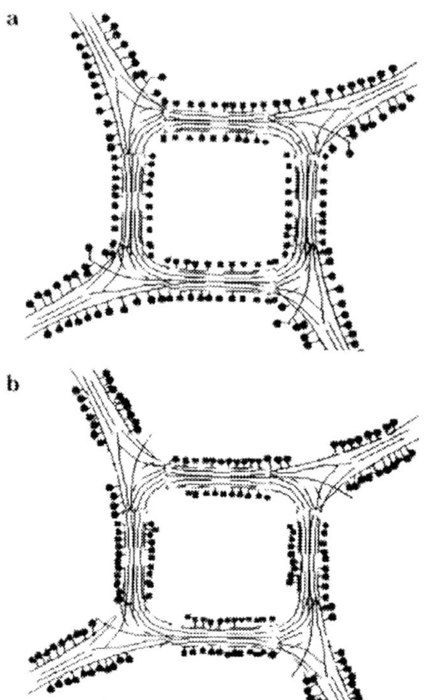

Figure 16. Schematic drawing of microstructure over the percolation threshold, made from Nafion polymer solutions (a) and a system in which no percolation takes place (b) (with permission from the American Chemical Society).

## 5.3. ION TRANSPORT IN PERFLUORINATED IONOMERS

Figure 17 shows the dependence of proton conductivities on the water content $\lambda$ for three kinds of perfluorinated ionomer membranes [9]. Proton conductivity strongly depends on the ion exchange capacity and water content in the membrane. This is because of the specific microscopic structure of the polymer, where the ion conduction path is determined by the channel structures of the membrane. This figure also implies the importance of the problem "water management in PEFC", because the amount of water in the membrane should be kept above a certain level by controlling the water flux in the membrane during the operation (see Section 6.4).

Ion conductivities of Nafion® membranes are measured for various cation forms. The mobilities of cations inside the membranes are plotted against the mobilities in the infinitely diluted aqueous solutions (figure 18) [57]. The mobility of cations inside the membrane is smaller than that in aqueous solutions by a factor of 1.3 to 4, indicating the limited freedom of motion in the channel, where cations migrate through restricted paths. In aqueous solutions on the other hand, ions move in three-dimensional space against the viscous force, which allows a larger freedom to ions. A remarkable difference in behavior is observed, i.e., the order of mobilities for different cationic species is reversed between the membrane and the solution phases. The reason of this different behaviors is explained when we take a role of water molecules around the ions into account.

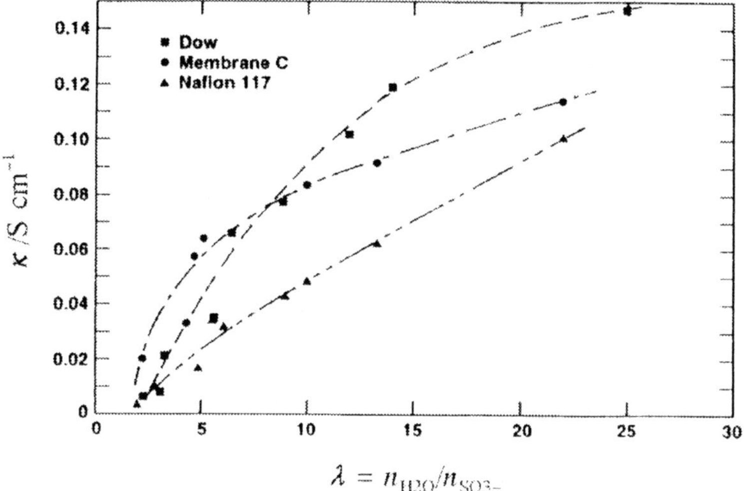


$$\lambda = n_{H2O}/n_{SO3^-}$$

Figure 17. Conductivity of PFSI membranes at 30 °C as a function of membrane water content (with permission from the Electrochemical Society).

The diameters of micropores in Nafion[®] 117 membranes are calculated by the fixed charge theory using electro-osmosis coefficients measured in the streaming potential [58]. Here it is assumed that water molecules move in the capillary pores inside the membrane, which is a hydrodynamic model of ionic channels. In figure 19, a plot of cationic mobility against the micropore cross section for several kinds of exchanged cations shows a simple relationship, and from this result it is inferred that the size of ionic channel is a crucial factor that determines the ionic mobility inside the channel.

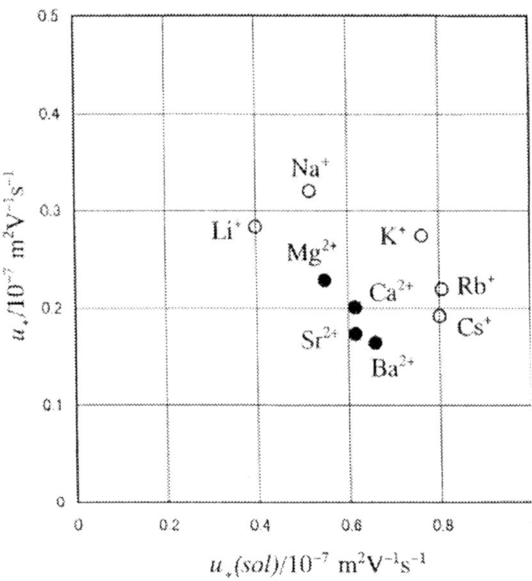


Figure 18. Cationic mobility inside Nafion® 117 membranes plotted against cationic mobility in aqueous solution (with permission from Elsevier Science).

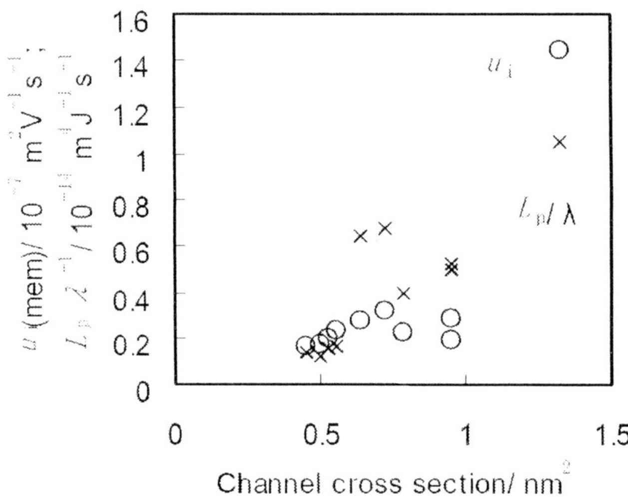


Figure 19. Cationic mobility $u_i$(mem) (○) and $L_p/\lambda$ (✕) in Nafion® 117 membranes plotted as a function of the channel cross section as calculated by fixed charge theory.

## 5.4. WATER TRANSPORT IN PERFLUORINATED IONOMERS

In figure 20, water transference coefficients $t_{H_2O}$ measured for various cation form Nafion® 115 membranes are plotted against the water content in the membrane [57]. According to the plots, cations can be categorized into three groups. For alkali or alkaline earth metal cations, $t_{H_2O}$ increases with cation hydrophilicity, and the membrane water content follows the same trend. When plotted against the ionic radius of cations, $t_{H_2O}$ increases with a decrease of the cationic radius. For hydrophobic cations like alkyl-ammonium cations, dependence of $t_{H_2O}$ on water content is very steep, and $t_{H_2O}$ increases with the radius of the cation. The H-form membranes are very unique, and $t_{H_2O}$ takes only small value of 2.6, although water content is very high.

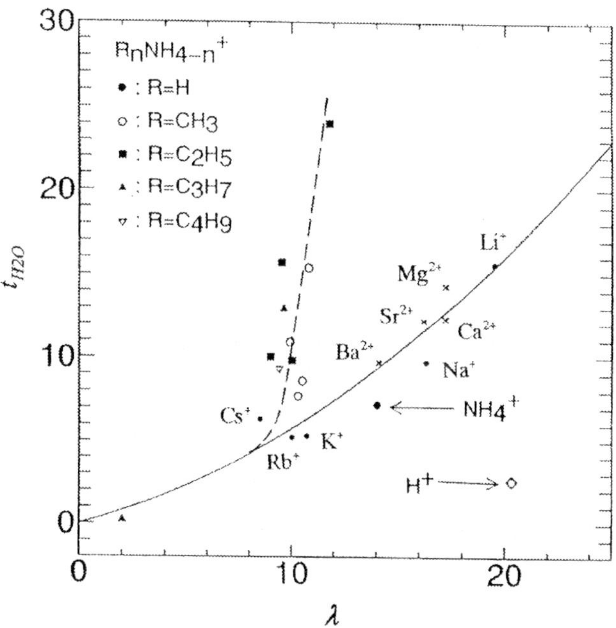


Figure 20. Water transference coefficients $t_{H_2O}$ measured for various cation form Nafion® 115 membranes plotted against the water content in the membrane (with permission from the Royal Society of Chemistry).

The water transference coefficient $t_{H_2O}$ for alkali and alkaline earth metal cations are correlated with the hydration enthalpy in water. Figure 21 depicts this relationship, and this correlation strongly indicates that water molecules coupled to ion transport consist in part of hydrated water molecules around the ion. It is interesting to observe that the number of water molecules carried by a cation, $zt_{H_2O}$, in some cases becomes larger than the number of hydrated water around cation in aqueous solutions. This needs to be explained by considering spatial circumstances around cations inside the channel structure of the membrane.

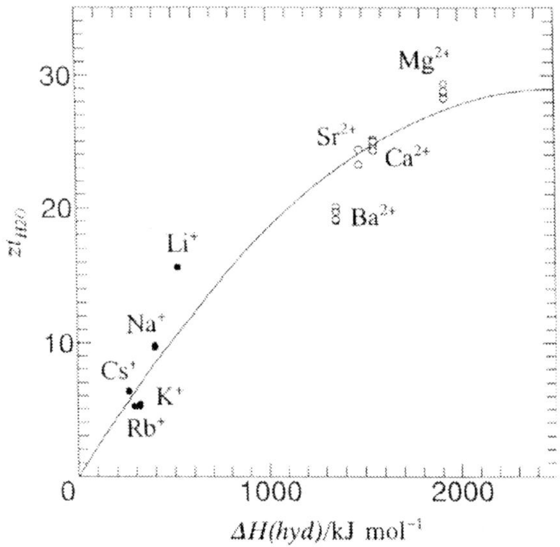


Figure 21. Water transference coefficient $t_{H_2O}$ multiplied by cationic valence $z$ in Nafion® 115 membranes of various cation forms plotted against the hydration enthalpies of cations (with permission from Elsevier Science).

It is assumed that ion transport is accompanied by water transport in two ways [59]. The first one is an electrostatic effect where charge-dipole interactions are dominating, and which keeps a certain amount of water located around a cation at any time. The second effect arises from ionic size, and water is "pushed" along the ionic channel when the cation is transported. The larger the cation, the more water is dragged by a volume exclusion effect. This simultaneous transport of ions and water molecules must occur along hydrophilic domains in the membrane. This situation is expressed by a simple relation [60]:

$$t_{H_2O} = t_{H_2O}(hyd) + t_{H_2O}(vol) \tag{42}$$

where $t_{H_2O}(hyd)$ is water of hydration that is Coulombically hydrated or peripherally bound to the cation, and $t_{H_2O}(vol)$ stands for the water beyond the hydrated water that accompanies the cation (pushed water by volume exclusion in the channel). The water of hydration for hydrophilic cations is assumed to be the same as that in the solution. Alkyl-ammonium ions, which usually have large volumes and low surface charge densities, do not actively interact with water and the surrounding sheath of water molecules will be very small. In this case neighboring water molecules are mostly pushed away from the ions in the channel.

A simple addition of two terms as expressed in Eq. (42) may be overestimation, but it is interesting to see that when pushed water $t_{H_2O}(vol)$ is plotted as a function of the Stokes radius of the ion (radius of the hydrated ion), a common curve appears for both the hydrophilic and hydrophobic cations [60]. In figure 22 (a), water transference coefficient $t_{H_2O}$ is plotted as a function of ionic radius, which is either crystallographic radius for alkali or alkaline earth metal cations or calculated ones from the partial molar volume for alkyl-ammonium cations assuming spherical shape [53]. Two groups of plots appear, and $t_{H_2O}$ decreases with ionic radius for alkali or alkaline earth metal cations while $t_{H_2O}$ increases with ionic radius for alkyl-ammonium cations. These opposite trend may arise from the different nature of two cation groups, and hydrophilic cations would carry water mostly with hydration sheath while alkyl-ammonium cations with hydrophobic skeletons would carry water mostly by volume exclusion effect. When the amount of water "pushed" by cations inside the ionic channel, $t_{H_2O}(vol)$, is calculated using Eq. (42) and plotted as a function of effective radius (hydrated radius for hydrophilic cations and that calculated from the partial molar volume for alkyl-ammonium cations), they merge into one unique line as seen in figure 22 (b). These results rationalize the hypothesis that cations move preferentially through hydrophilic domains in the membrane, and that hydrophilic cations hydrate as if they were in aqueous solutions, carrying additional water by volume exclusion in the narrow channel structure.

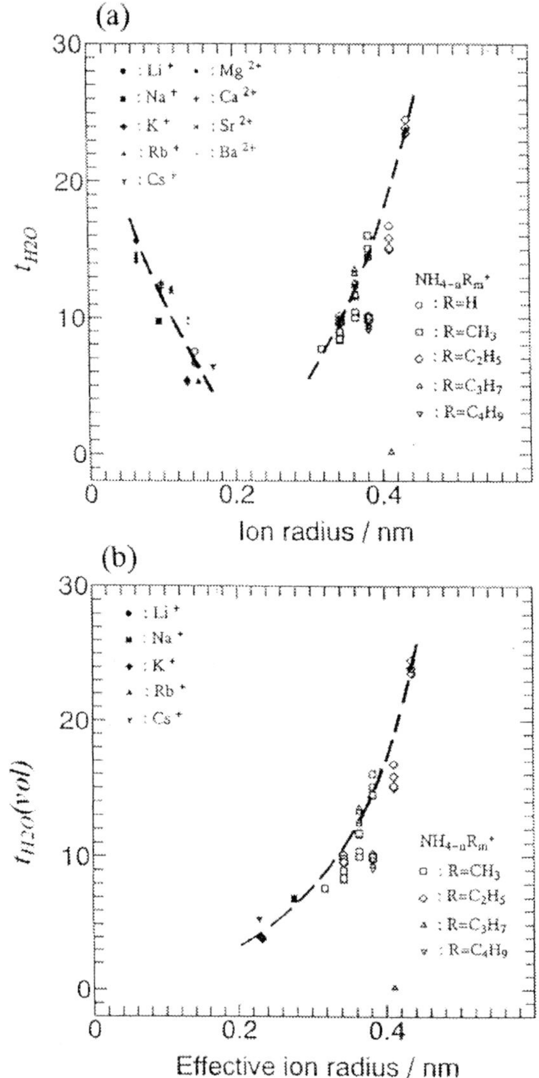



Figure 22. (a) $t_{H_2O}$ as a function of ionic radius. (b) $t_{H_2O}(vol)$ as a function of effective ionic radius (with permission from Elsevier Science).

The relation between ion and water transport in the ionic channel is viewed from another point. When the membrane specific resistance $\kappa^{-1}$ is plotted against the water transference coefficient, two categories are observed for the cations (figure 23). For the bulky alkyl-ammonium cations, the membrane resistance increases as the water transference coefficient, $t_{H_2O}$, increases, which is the size effect in $t_{H_2O}$. For hydrophilic cations, on the other hand, $\kappa^{-1}$ versus $t_{H_2O}$ curve shows a maximum around $t_{H_2O} = 10$, and after that $\kappa^{-1}$ slightly decreases with $t_{H_2O}$. The hydrophilic cations like $H^+$ and alkali or alkaline earth metal cations, bring more water into the membrane than the alkyl-ammonium cations (for which $\lambda$ is almost constant around 10), and water molecules can enlarge the ionic channel resulting in the lower resistance in the membrane. It is natural that alkyl-ammonium cations of larger size migrate more slowly than hydrophilic cations of smaller size, but there is more to it that explains the role of water molecules inside the channel.

The water transference coefficients of alkyl-ammonium cations are larger than, or at least close to, the water content of the membranes (which are all close to 10), as can be seen in figure 23. Unlike the case of alkali and alkaline earth metal cations, $t_{H_2O}$ is predominantly determined by their volume or by the membrane water content. These hydrophobic cations exclude water molecules from the membrane. This indicates that only small fraction of the ion exchange sites seem to participate in the formation of ion-water clusters while the rest of the exchange sites are inactive and cannot contribute in ionic conductance of the membrane. The high membrane resistance observed for large size alkyl-ammonium ions would thus be caused not only by the plugging effect but also by the water exclusion effect, where a strong ion-pairing occurs between cations and ion exchange sites.

The water permeabilities $L_p$ in Nafion® 117 membranes of various cation forms are calculated by Eq. (23) and plotted in figure 24 as a function of water content [57]. This parameter is defined as a measure of water flow by pressure difference across the membrane, and is related to water diffusion coefficient (pressure diffusion or normal diffusion). Because higher water content means larger domain of hydrophilic paths in the ionic channel, this result shows water diffusivity is related to the size of the ionic channel. In figure 19, a parallel relationship is observed between $L_p/\lambda$ and the channel cross section, which also supports this assumption. A very simple picture of ionic channel is conjectured: transport of species in the membrane, whether cations or water molecules, the latter being either dragged by ions or by free diffusion, are all controlled by the size of ionic channels inside the membrane. A failure from this rule is $H^+$, and it shows a higher conductivity than expected from the trend of the other cations. Also less water is coupled to $H^+$ in the electro-osmotic drag. $H^+$ will move by

Grotthuss mechanism (hopping mechanism) in the channel structure while normal cations move by vehicle mechanism, like in aqueous solutions [40].

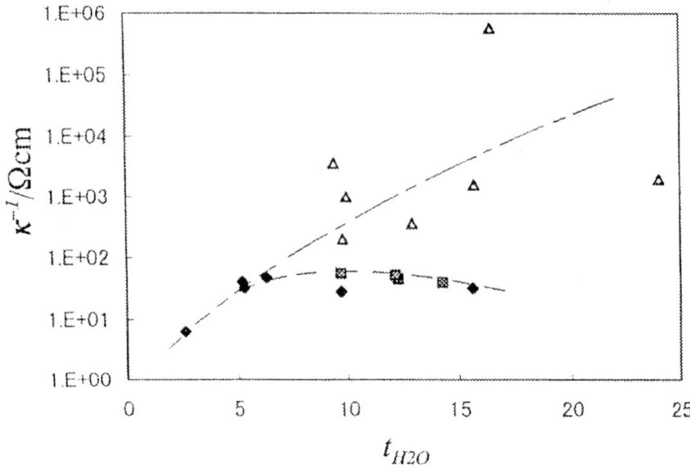

Figure 23. Membrane specific resistance $\kappa^{-1}$ plotted against the water transference coefficient $t_{H_2O}$ in Nafion 117® membranes of various cation forms. $\triangle$: alkyl-ammonium cations, $\blacklozenge$: H⁺ and alkali metal cations, ■: alkaline earth metal cations (with permission from Elsevier Science).

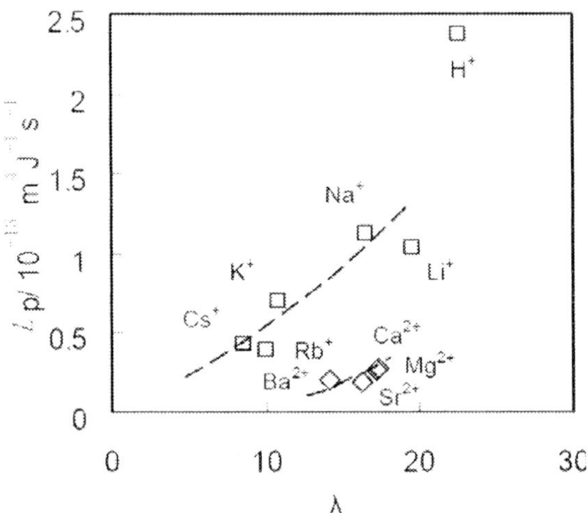


Figure 24. Water permeability $L_p$ in Nafion® 117 membranes of various cation forms plotted as a function of water content.

## 5.5. STATE OF WATER IN THE MEMBRANE

In perfluorinated ionomer membranes, the properties and performances are directly related to their morphology, and the supramolecular organization of ionic and crystalline domains may change by the processing history. In this context, the state of water inside the membrane should also affect the atmosphere in ionic channels. Because of the lowered surface charge density of sulfonic acid groups by the shielding effect of fluorinated main chains, hydrophilicity of counter-cations may influence the water content in the membrane. The entropy change due to release of water becomes relatively large during cation exchange processes [61]. In this context the state of water molecules is worth investigated using several techniques.

DSC curves on Nafion® membranes of various cation forms are summarized in figure 25. From these data different states of water in the equilibrium state of the membrane are categorized in table 2, as the number of water molecules per cation exchange site [39]. The first type of water is more bulk-like than the other water, and it is defined as "semi-free water", because this water freezes around – 20 °C when the membrane samples are cooled from room temperature [62]. The second type of water does not freeze down to –120 °C, and is called "bound water", because this water is likely to be in strong interaction with either the cations or sulfonic acid groups [14], or confined in the transition region between hydrophilic and hydrophobic domains of the membrane (with disordered structure), and is very different from the bulk-like water.

**Table 2. Results of DSC and the streaming potential measurements for the Nafion® 117 membranes of various cation forms**

| Membrane form | HM | LiM | NaM | KM | RbM | CsM | MgM$_2$ | CaM$_2$ | SrM$_2$ | BaM$_2$ |
|---|---|---|---|---|---|---|---|---|---|---|
| Water content $\lambda$ | 22.6 | 19.5 | 16.5 | 10.8 | 10.1 | 8.6 | 17.3 | 17.2 | 16.2 | 14.1 |
| Freezing water/cationic site | 9.9-10.6 | 8.7-9.3 | 7.4-7.8 | 4.3-4.5 | 2.9-3.1 | 3.3-3.5 | 8.3-8.9 | 7.5-8.0 | 7.2-7.6 | 4.3-4.6 |
| $t_{H_2O}$ | 2.6 | 15.6 | 9.7 | 5.3 | 5.2 | 6.3 | 14.3 | 12.3 | 12.2 | 9.7 |
| Ratio freezing water/$\lambda$ | 0.44-0.47 | 0.45-0.48 | 0.45-0.47 | 0.40-0.42 | 0.29-0.31 | 0.38-0.41 | 0.48-0.51 | 0.44-0.47 | 0.44-0.47 | 0.30-0.33 |

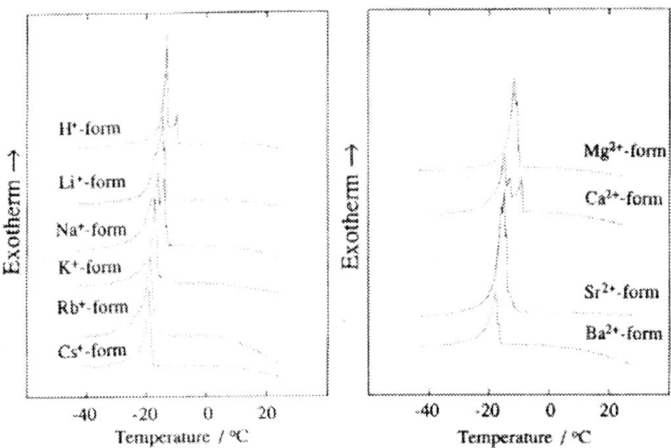



Figure 25. DSC curves of Nafion$^{®}$ 117 membranes of $H^+$ and alkali metal cation forms (with permission from the Electrochemical Society of Japan).

In figure 26 the number of freezing water (semi-free water per cationic site) in DSC is plotted against the number of water molecules "pushed" by hydrated cations. Interestingly, there appears a correlation between two parameters, and it is inferred that actually these semi-free water may be the water molecules pushed by volume exclusion by moving cations inside the ionic channels.

The amount of semi-free water is about 50% of the total water in the membrane, the trend being that it decreases as the hydrophilicity of the cation decreases. Total amount of water also decreases and the DSC peak shifts to more negative temperatures, suggesting that the water becomes less bulk-like and more confined in the polymer structure when it depletes. FT-IR spectra of recast Nafion films of various cation forms are presented in figure 27 [44]. In figure 28, the intensities and wave numbers of OH stretching and bending signals are shown as a function of water content. Less hydrophilic cations result in a smaller extent of hydrogen bonding and less water signals. Although DSC categorized non-freezing water as the bound water to cations or sulfonic acid groups or those residing in the interphase regions near hydrophobic domains, the IR could not discriminate them.

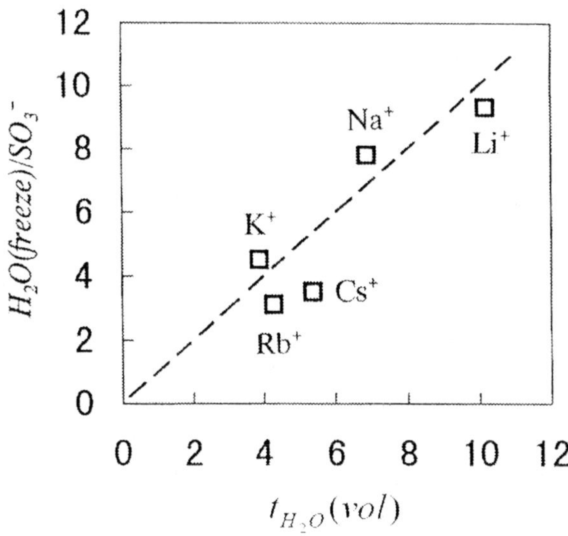

Figure 26. Number of freezing water molecules per cation exchange site plotted against the number of water molecules "pushed" by hydrated cations in Nafion® membranes (with permission from Elsevier Science).

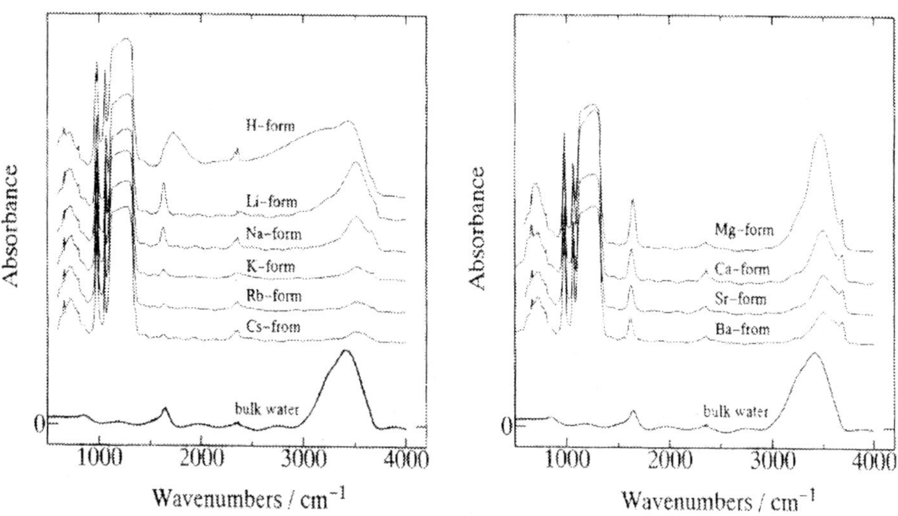

Figure 27. FT-IR spectra of recast Nafion films of various cation forms (with permission from Oldenbourg Wissenschaftsverlag).

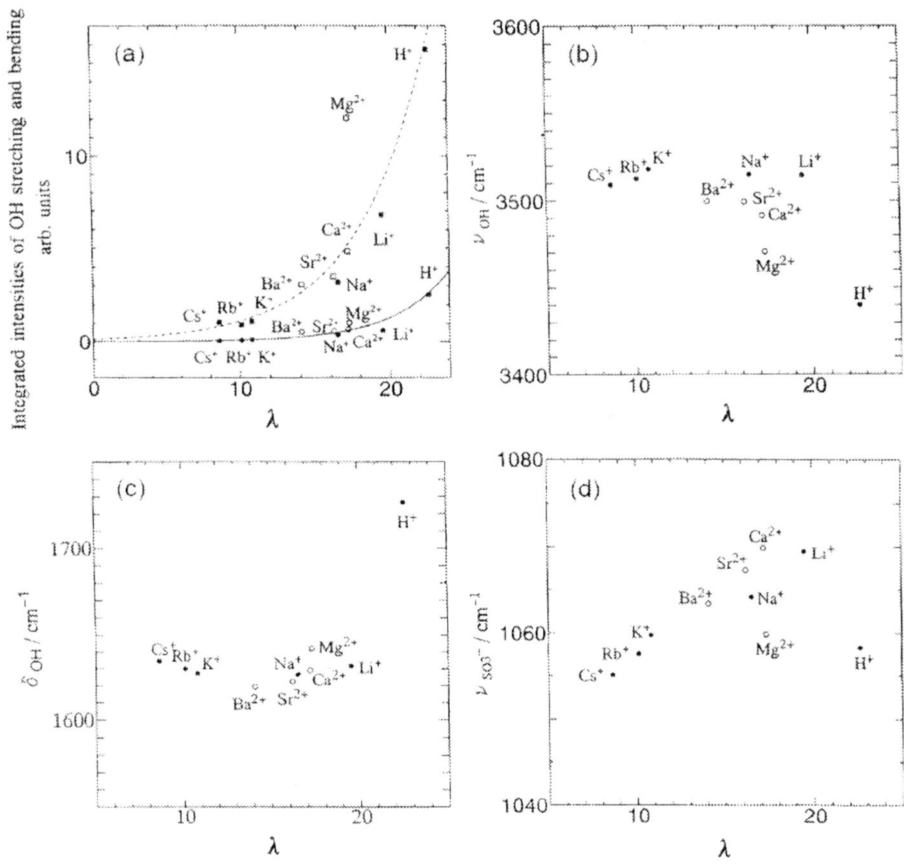

Figure 28. Intensities (a) and wave numbers of OH stretching (b) and bending (c) signals, and wave numbers of symmetric stretching of $SO_3^-$ (d), as a function of water content (with permission from Oldenbourg Wissenschaftsverlag).

The peak position of the symmetric stretching of $SO_3^-$, $\nu_{SO_3^-}$, shows a clear tendency, and the interactions between the cations and the sulfonic acid groups in fully hydrated Nafion membranes follows the order: Li > Na > K > Rb > Cs, and Ca > Sr > Ba > Mg for alkali and alkaline earth metal cation form membranes, respectively (figure 28(d)) [44]. This means that the polarization of S-O dipole is stronger in this order, which appears to be consistent with the hydrogen bonding as observed by OH stretching and bending modes. The image of the simple charge-charge interactions between cations and $SO_3^-$ sites should thus need careful consideration, because in this image more hydrophilic cations with larger hydration radius should have resulted in smaller interaction and less positive shift in the $V_{SO_3^-}$ , which is against the experimental result.

# 5.6. DIFFUSION COEFFICIENTS OF ION AND WATER: EFFECT OF ION EXCHANGE CAPACITY

Mass transport of proton and water molecule in PFSI membranes is influenced by the polymer structures. In this section, the relationships between the polymer structure (especially the equivalent weight (*EW*) value) and mass transport (proton and water molecule) are discussed with three PFSI membranes. Nafion[®] and two kinds of commercial membranes (PFSI (A) and PFSI (B)) with variety of *EW* values are evaluated especially by pulsed-field-gradient spin-echo NMR (PGSE-NMR), which is reported to be a powerful tool for the measurement of diffusion coefficients of species in the membrane [63-65].

Figure 29 shows the water volume fraction $\theta$, water content $\lambda$ and membrane density $d_{wet}$ for the membranes in the fully hydrated state [66]. The membranes swell by absorption of water molecules especially for the smaller *EW* membranes. It is well known that the PFSI membranes absorb water molecules to form ion cluster regions together with sulfonic acid groups and protons [2,4,27]. The protons dissociated from the sulfonic acid groups transport through the ion cluster regions in the medium of water molecules. As the *EW* decreases (ion exchange capacity increases), the volume fraction of water (volume of ion cluster regions) tends to increase.

Figures 30 and 31 depict the proton conductivity $\kappa$ and two water mobility parameters (water transference coefficient $t_{H_2O}$ and water permeability $L_p$) for three PFSI membranes with different *EW* values [66]. Both $\kappa$ and $L_p$ values for the H-form membranes increase with decreasing the *EW* value of the membranes. Together with figure 29, this indicates that larger membrane expansion by water swelling promotes transport of proton and water molecules due to the expanded ion cluster regions. In addition, $\kappa$ and $L_p$ values for the H-form membranes are quite higher than those of the other ion form membranes, while $t_{H_2O}$ values exhibit the reverse trend. This means that in fully hydrated PFSI membranes the proton moves by Grotthuss mechanism [67] not by vehicle mechanism [34], together with water molecules as observed in Li and Na-form membranes. In fact, the mobilities $u$ of proton is quite higher than those of the other ions in spite of almost the same concentration $C$ of carrier ions (figure 32). Zawodzinski et al.

demonstrated with PGSE-NMR method [63], that proton transport by Grotthuss mechanism was accelerated with increasing water content of Nafion® membranes.

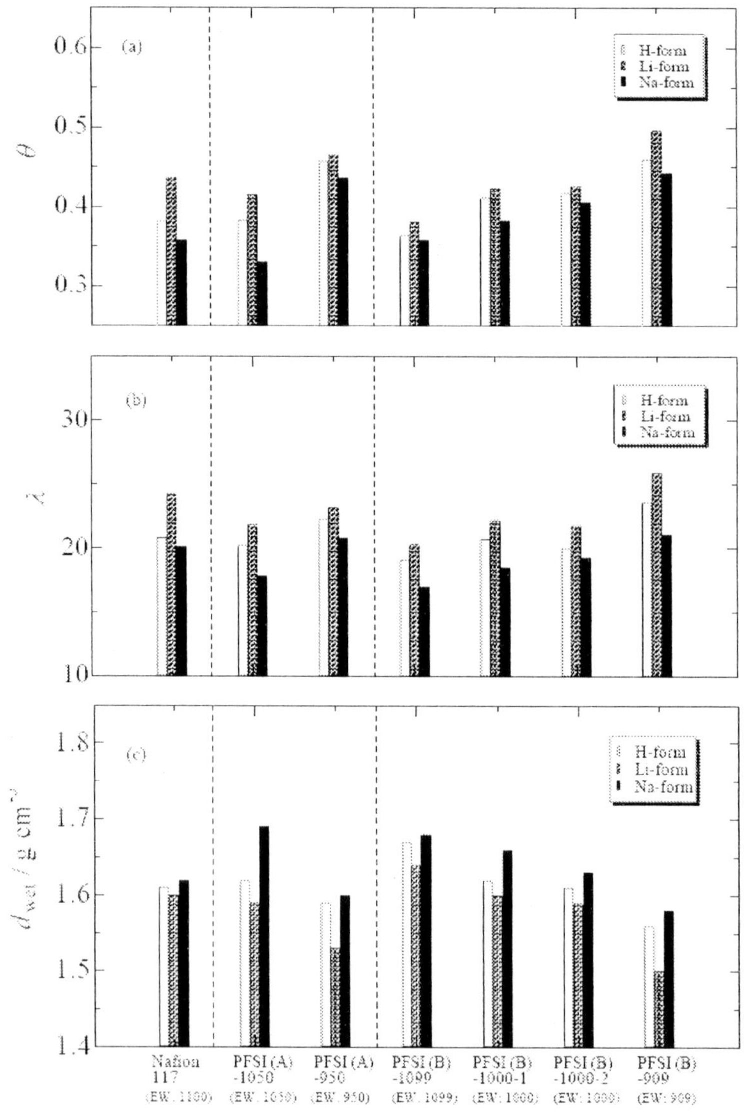

Figure 29. (a) Water volume fraction $\theta$, (b) water content $\lambda \equiv n_{H_2O}/n_{SO_3^-}$ and (c) density $d_{wet}$ for the 7 different membranes in the fully hydrated state (with permission from the American Chemical Society).

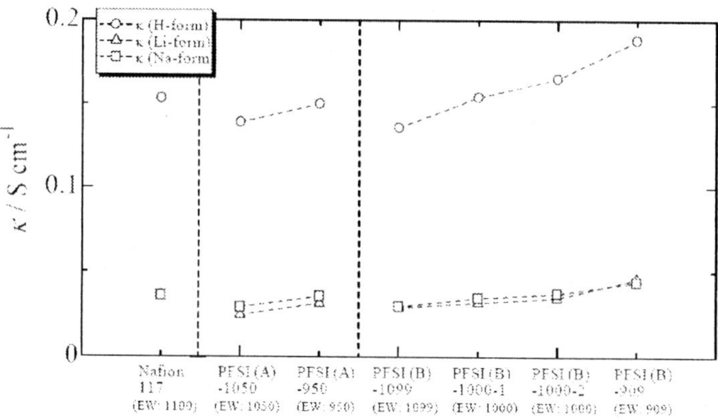

Figure 30. Ionic conductivity $\kappa$ for the 7 different membranes in the fully hydrated state (with permission from the American Chemical Society).

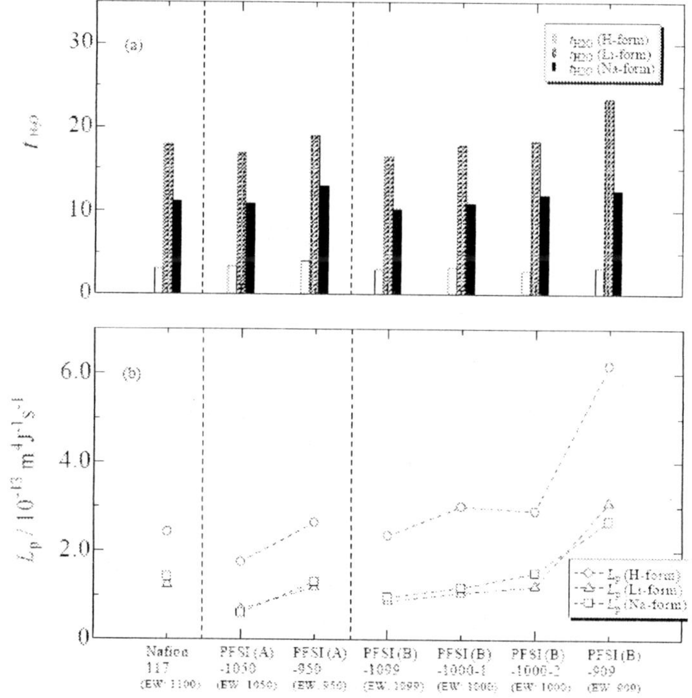

Figure 31. (a) Water transference coefficient $t_{H_2O}$ and (b) water permeability $L_p$ for the 7 different membranes in the fully hydrated state (with permission from the American Chemical Society).

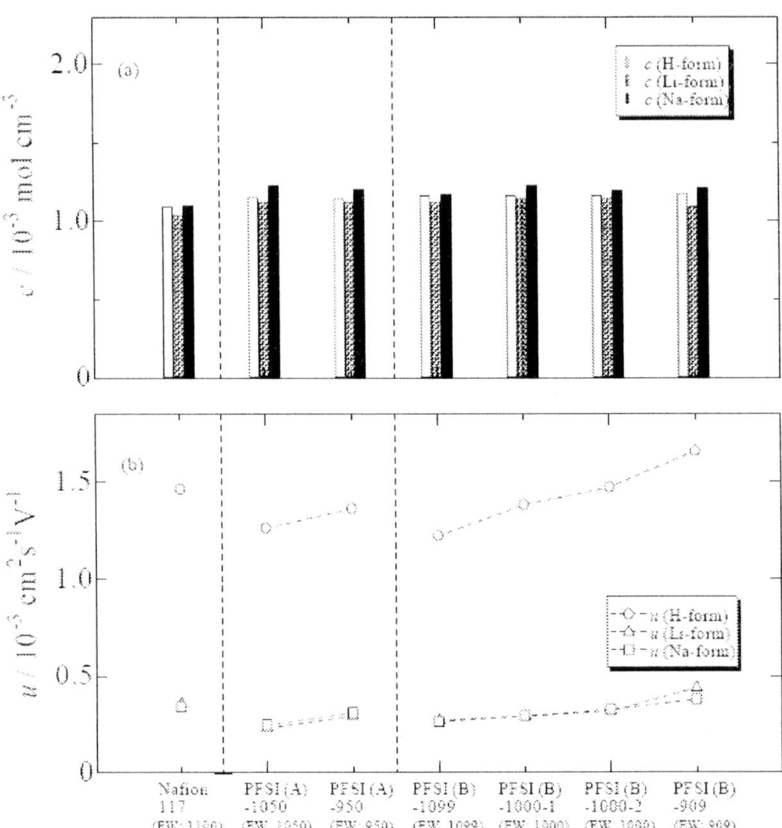

Figure 32. (a) Concentration $c$ and (b) mobility $u$ of carrier cation species in the 7 different membranes in the fully hydrated state (with permission from the American Chemical Society).

The correlation between proton and water molecule transports in the PFSI membranes is quite interesting also for practical use, and this fact suggests the importance of water management in the PEFC systems to suppress the performance deterioration. To further study on the mass transport behaviors, PGSE-NMR method is applied to determine the self-diffusion coefficient of each species containing magnetically active nuclei in the polymer electrolyte. Figure 33 shows the self-diffusion coefficients of water molecule $D_{H_2O}$ in the PFSI membranes with different $EW$ values [66]. The trend of $D_{H_2O}$ against $EW$ is in good agreement with that of the $L_p$ value (figure 31). For the Li-form membranes the self-diffusion coefficient of $Li^+$ ion, $D_{Li^+}$ is also evaluated (figure 34). The $D_{Li^+}$ value is approximately one forth of the $D_{H_2O}$ value and increases with decreasing the $EW$ value. This suggests that the transport rate of $Li^+$ ion in the PFSI membranes is correlated with that of water molecule.

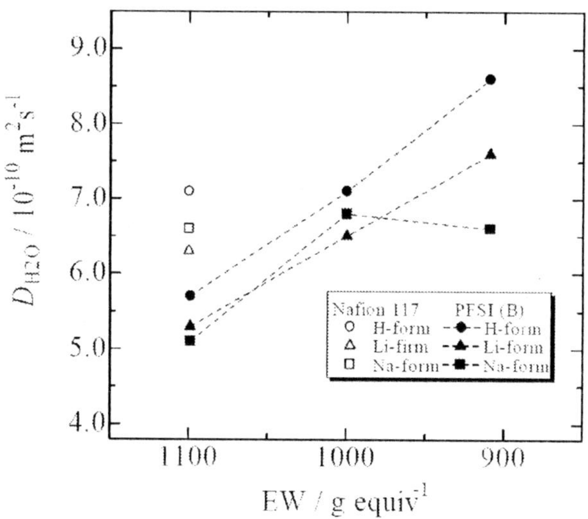


Figure 33. Self-diffusion coefficient of water molecule $D_{H_2O}$ in the membranes against $EW$ value. PFSI (B)-1000-2 was used for the measurements on PFSI (B)-1000 (with permission from the American Chemical Society).

Thermal property of the membranes, especially at low temperatures, is one of the critical factors for practical use of PEFCs systems, because of the freezing problem of water molecules in the membrane. This phenomenon causes a drastic reduction of proton conductivities in the membrane [14], lowering the PEFC performances. The PGSE-NMR method is useful to examine the proton mobility in the lower temperatures because of the easy operation in changing the measurement temperature. Figure 35 shows the Arrhenius plots of the $D_{H_2O}$ for different cation form PFSI membranes in the temperature range of 30 to -40 °C [68]. All of the plots follow similar trends from 30 to -15 °C, and the $D_{H_2O}$ values in the membranes are about three to five times lower than that in pure water. Discontinuous decrease in the $D_{H_2O}$ value at -15°C and following steeper temperature dependence are clearly observed. Comparison of the slope of curves infers that water in the membranes diffuses like bulk water above -15 °C, but the mobility is limited. The water in the membranes is partially frozen around -20 °C [39], but non-freezing water exists in the membranes and moves below the temperature. Cappadonia et al. studied the temperature dependence of proton conductivity in Nafion® 117 membranes between 27 and -133 °C, where the membrane samples were pretreated at various conditions [69]. They observed different Arrhenius plots above and below around -13 °C, where the discontinuity occurred for all the membranes. These tendencies are in good agreement with the present results.

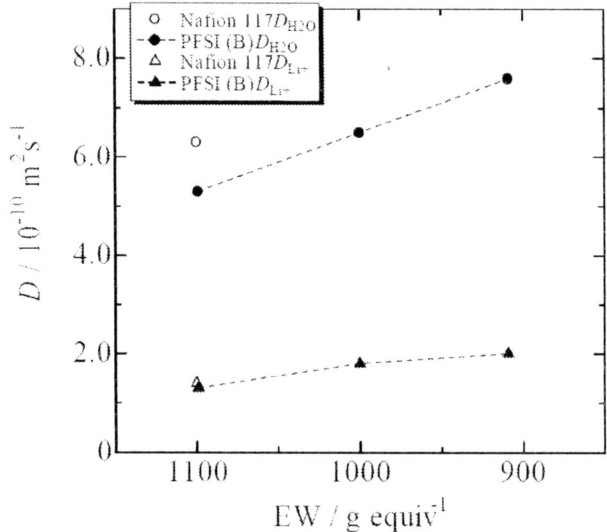


Figure 34. Self-diffusion coefficients of water molecule $D_{H_2O}$ and Li$^+$ cation $D_{Li^+}$ in the Li-form membranes against $EW$ value. PFSI (B)-1000-2 was used for the measurements on PFSI (B)-1000 (with permission from the American Chemical Society).

Figure 36 shows the DSC curves for the membranes in the cooling process. For all the membranes, no exothermic peaks are observed between 50 and -15 °C, and around -20 °C peaks due to the freezing of water appears. This is in good agreement with the temperature dependent behaviors of $D_{H_2O}$ in the membranes. The fact that peaks split, suggests the existence of various states of water in the ion channels [39]. Table 3 summarizes the amount of freezing and non-freezing water estimated from the freezing peak area in the DSC curves [68]. For all of the

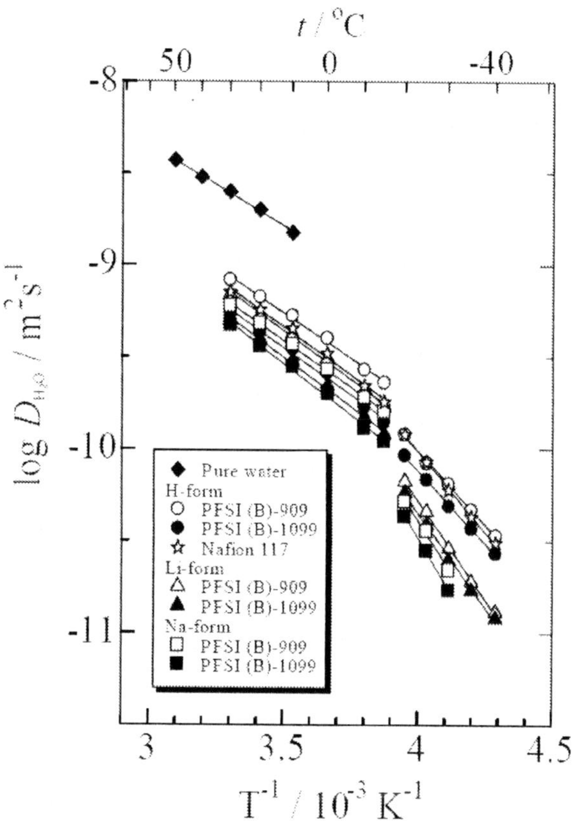


Figure 35. Arrhenius plots for water diffusion coefficients $D_{H_2O}$ of PFSI (B) X-form membranes having different $EW$ values along with Nafion® 117 H-form membrane (with permission from the American Chemical Society).

membranes, the number of freezing water molecules increases with decreasing $EW$ value independent of the cation types. This fact accords with the trend that $D_{H_2O}$ increases with decreasing $EW$ value, because the freezing water corresponds to weakly interacting water with ionic species in the channels as compared with non-freezing water. In the Li and Na-form membranes the number of non-freezing water molecules is almost $EW$ independent, although the number of freezing water molecules increases with decreasing $EW$ value. This implies that the number of non-freezing water mainly depends on the cation species and membrane types, while the number of freezing water depends on the $EW$ value. On the other hand, in the H-form membranes, both numbers of freezing and non-freezing water molecules increase with decreasing *EW* value but the average ratio is almost constant. This indicates that proton in the membranes exists as an oxonium ion $H_3O^+$ and the rate of proton exchange with $H_2O$ is quite fast and undistinguishable, which also supports that proton moves fast mainly by Grotthuss mechanism in the membranes.

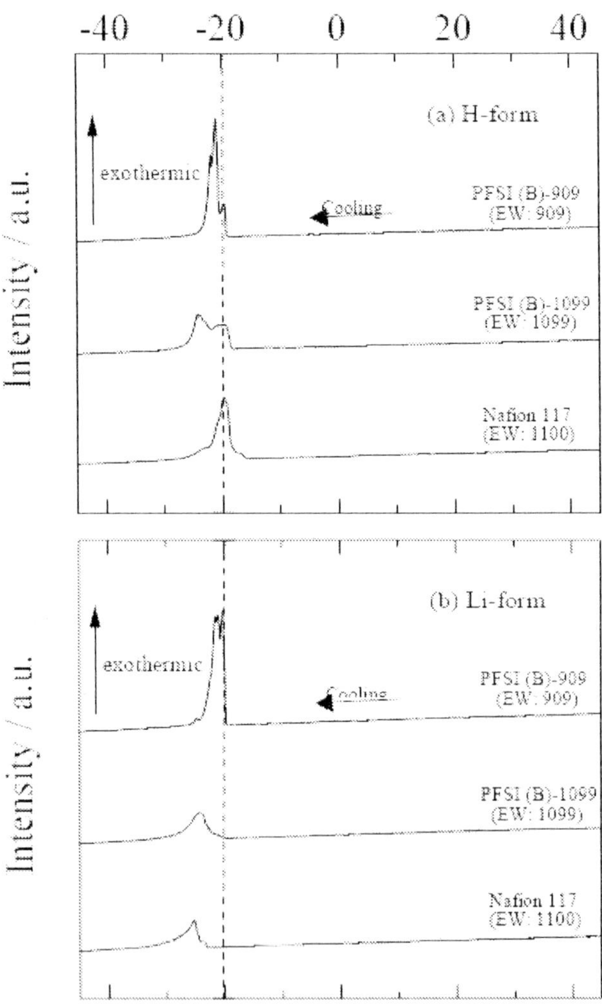


Figure 36. Continued on next page.

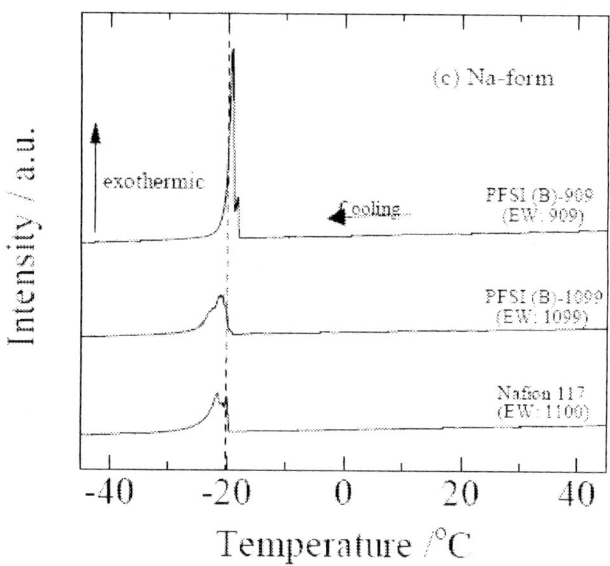


Figure 36. DSC curves of PFSI (B) X-form membranes having different EW values along with Nafion 117 X-form membrane in the fully hydrated state measured in the temperature decreasing process (with permission from the American Chemical Society).

## Table 3. Water conditions in PFSI (B) membranes along with Nafion® 117 in the fully hydrated state (with permission from the American Chemical Society)

| Membrane | PFSI (B)-909 ($EW$: 909) | | | PFSI (B)-1099 ($EW$: 1099) | | | Nafion® 117 ($EW$: 1100) | | |
|---|---|---|---|---|---|---|---|---|---|
| Cation form | H | Li | Na | H | Li | Na | H | Li | Na |
| Freezing enthalpy [a] /J g$^{-1}$ | 37.0 | 47.1 | 44.7 | 29.9 | 22.0 | 26.6 | 30.2 | 19.4 | 29.4 |
| Water content $\lambda$ [b, c] | 23.6 | 25.9 | 21.1 | 19.1 | 20.3 | 17.0 | 20.8 | 24.3 | 20.1 |
| Freezing water [c] | 8.3-8.9 | 10.8-11.6 | 9.8-10.5 | 7.2-7.7 | 5.4-5.8 | 6.3-6.8 | 7.7-8.3 | 5.0-5.4 | 7.3-7.8 |
| Non-freezing water [c] | 14.7-15.3 | 14.3-15.1 | 10.6-11.3 | 11.4-11.9 | 14.5-14.9 | 10.2-10.7 | 12.5-13.1 | 18.9-19.3 | 12.3-12.8 |
| Average ratio of freezing water | 0.37 | 0.43 | 0.48 | 0.39 | 0.27 | 0.39 | 0.39 | 0.21 | 0.37 |

[a] calculated from the area of DSC peaks for the water freezing.
[b] Reference 66
[c] number of water molecules per a cation exchange site.

## 5.7. ION AND WATER TRANSPORT IN BINARY CATION SYSTEMS

When two cations coexist in the membrane, a strong interference is expected during transport of these cations. When two kinds of cations transport by way of the same ion exchange sites, a strong Coulombic interaction may arise among them. The experimental result actually depends on the selection of cation species. In cation mixtures $H^+/Na^+$ and $H^+/Ca^{2+}$, the membrane conductivity changed almost linearly with the cation composition in the membrane. Figure 37 depicts the cation mobilities against cationic composition in H/Na and H/Ca mixed form Nafion® 117 membranes, which shows mobilities of two cations keep at all times constant [57]. This fact is remarkable, considering the high concentration of cation exchange sites in the membrane (1-2 mol $dm^{-3}$), and the membrane preference for one kind of cation to the other in ion exchange isotherms (see section 5.8).

This apparent contradiction seems to be resolved if we take into consideration the shielding effect on the fixed ion exchange sites by water and by cation association. Such a shielding has been reported for water of hydration around ions in the aqueous solution. However, further study discloses that this is not always the case for other cations. For example, for $H^+/Fe^{3+}$, $H^+/Ni^{2+}$ and $H^+/Cu^{2+}$ cation pairs, strong interferences are observed and $H^+$ mobility changes with cation composition depending on which cation coexists in the membrane [42]. Figure 38 depicts the mobility changes for these cation systems, and $H^+$ mobility decreases in the presence of $Fe^{3+}$, but increases in the presence of $Cu^{2+}$. $Ni^{2+}$ falls on the intermediate case, and very little interaction is observed between $H^+$ and $Ni^{2+}$ in the membrane. This difference is attributed to the nature of water molecules around ions, i.e., the self-exchange rate of water $k_w$ around cations increases in the order: $Fe^{3+}$ ($k_w \cong 10^2$ $s^{-1}$) < $Ni^{2+}$ ($k_w \cong 10^4$ $s^{-1}$) < $Cu^{2+}$ ($k_w \cong 10^8$ $s^{-1}$) [70]. It is anticipated that $H^+$ hopping would be favored around cations, which have a high exchange rate of hydration water molecules.

A non-linear relation is found between the water transference coefficient and the ionic transference number in Nafion® membranes in almost all the cation systems tested (see figure 39, for example). Deviation from the straight line indicates that the amount of water accompanying one specific ion is not constant, and water molecules interfere when water is dragged along with cations in mixed cation systems. Since cations in general do not interfere strongly in their movement, this interference might arise among the water molecules "pushed" in the ionic channel. Depending on the combinations of cation pairs, mobile water molecules are lost ($H^+/Na^+$, $H^+/K^+$, $H^+/Rb^+$, $H^+/Cs^+$, $H^+/Ca^{2+}$) or gained ($H^+/Li^+$, $H^+/Ni^{2+}$, $H^+/Cu^{2+}$, $Li^+/K^+$, $Li^+/Rb^+$) during the course of binary cation movement [37,41,42,57,71,72].

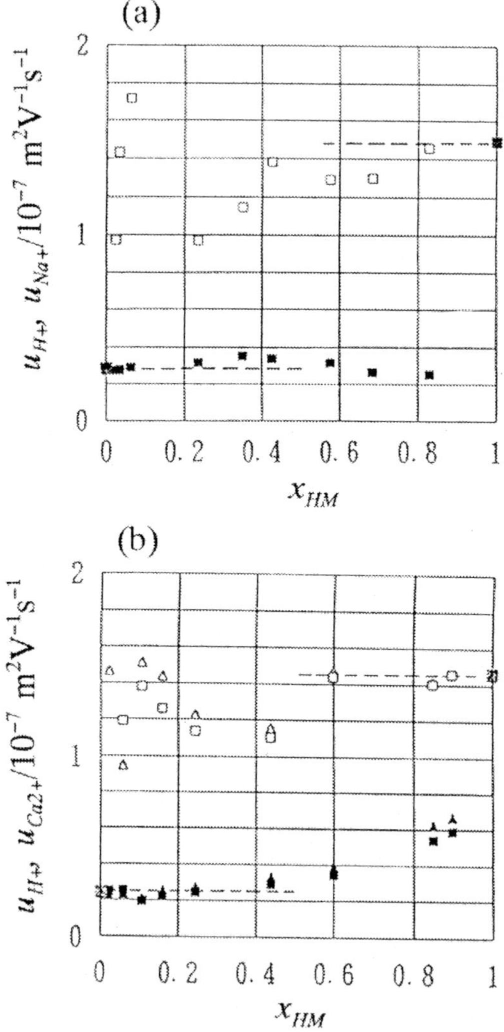

Figure 37. Change of cation mobilities with cationic composition $x_{HM}$ in H/Na and H/Ca mixed form Nafion® 117 membranes. $\triangle$, $\square$: $u_{H^+}$; $\blacktriangle$, $\blacksquare$: $u_{Na^+}$ and $u_{Ca^{2+}}$ (with permission from Elsevier Science).

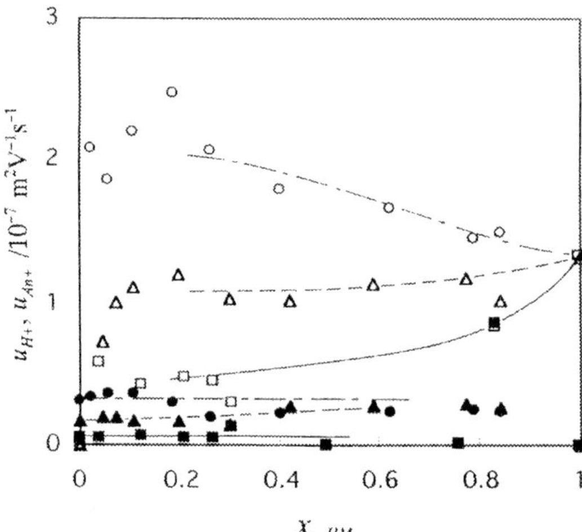

Figure 38. Change of cation mobilities with cationic composition $x_{\mathrm{HM}}$. ($\square$, $\blacksquare$), $u_{H^+}$ and $u_{Fe^{3+}}$ in the H/Fe system; ($\triangle$, $\blacktriangle$), $u_{H^+}$ and $u_{Ni^{2+}}$ in the H/Ni system; ($\bigcirc$, $\bullet$), $u_{H^+}$ and $u_{Cu^{2+}}$ in the H/Cu system (with permission from the American Chemical Society).

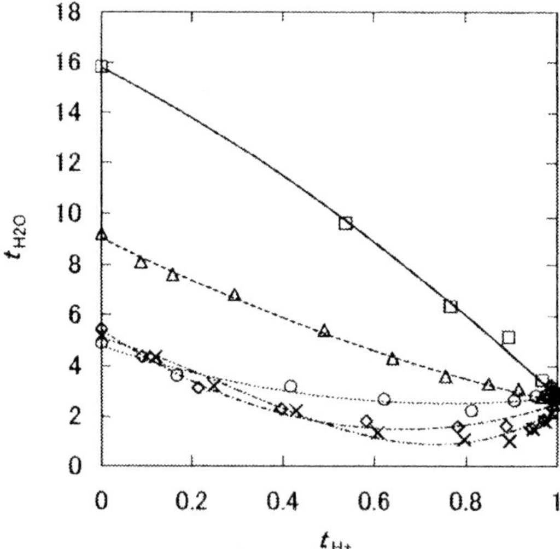

Figure 39. Water transference coefficient $t_{H_2O}$ in Nafion 115$^{\circledR}$ membranes of various cation forms plotted against ionic transference number of H$^+$ $t_{H^+}$ in Nafion$^{\circledR}$ membranes. $\square$, H/Li; $\triangle$, H/Na; $\circ$, H/K; $\diamondsuit$, H/Rb; $\times$, H/Cs systems (with permission from the American Chemical Society).

## 5.8. MOLECULAR INTERACTIONS – BINARY CATION SYSTEMS

Figure 40 depicts partition isotherms of several cations between the membrane phase and the surrounding aqueous solution phase, for binary cation systems in Nafion 117® membranes [41,42,71]. Except for $Li^+$, almost all the cations have higher exchange preference over $H^+$ with sulfonic acid groups in the membrane. Increasing the cation valence give stronger tendencies to be incorporated in the membrane. For alkali metal cations, higher atomic number resulted in higher affinity in the membrane. Higher affinity suggests stronger interactions between cations and the sulfonic acid groups, which infers that cation mobilities may decrease in this order.

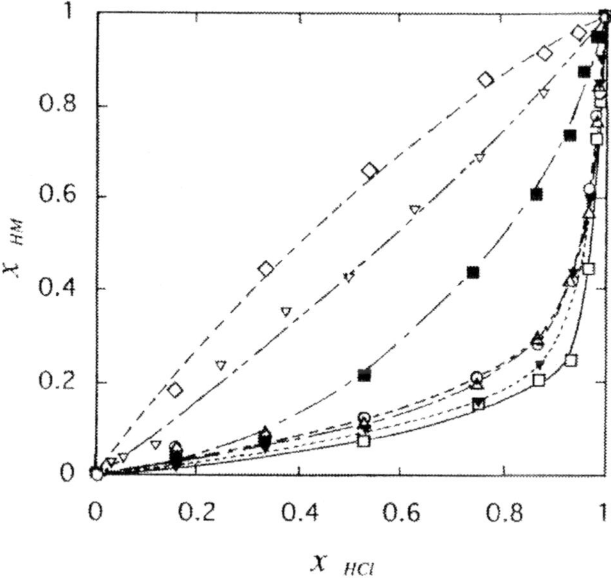

Figure 40. Cationic composition in the membrane $x_{HM}$ against solution composition $x_{HCl}$ at 25 °C for various kinds of binary cation systems. ◇, H/Li; ▽, H/Na; ■, H/K; ▼, H/Ca; □, H/Fe; △, H/Ni; ○, H/Cu (with permission from the Electrochemical Society, Elsevier Science, and the American Chemical Society).

The thermodynamic equilibrium constant $K_{th}$ is obtained from the equilibrium constant $K_{ex}$ of the exchange reactions for various cation pairs [43].

$$K_{th} = \frac{a_{ACl}a_{HM}}{a_{HCl}a_{AM}} = K'\frac{\gamma_{HM}}{\gamma_{AM}} \tag{43.1}$$

$$K' \equiv \frac{a_{ACl}x_{HM}}{a_{HCl}x_{AM}} = \frac{\gamma_{ACl}}{\gamma_{HCl}}K_{ex} \tag{43.2}$$

where $\gamma_{ACl}$ and $\gamma_{HCl}$ are activity coefficients of HCl and ACl in the solution, respectively, and are calculated using Debye-Hückel approximation. The Gibbs energy change for the exchange reaction is given as follows, using the chemical potential of relevant species:

$$\Delta G = \mu_{HM} + \mu_{ACl} - \mu_{HCl} - \mu_{AM} \tag{44}$$

which is 0 at the equilibrium state. Consider the standard state, where the activity coefficients of all the species are 1 and $\mu_i$ becomes $\mu_i^o$. In this case $\Delta G$ becomes $\Delta G^0 = -RT \ln K_{th}$, and

$$\Delta\mu_{HM} + \Delta\mu_{ACl} - \Delta\mu_{HCl} - \Delta\mu_{AM} = RT \ln K_{th} \tag{45}$$

where $\Delta\mu_i \equiv \mu_i - \mu_i^0 = RT \ln a_i$. Suppose $x_{HM}$ equivalent of HM and $x_{AM}$ equivalent of AM are brought together to make a mixture of H/A-form membrane. The Gibbs energy change of mixing $\Delta G_{mix}$ follows [43,73]:

$$\Delta G_{mix} = x_{HM}\Delta\mu_{HM} + (1 - x_{HM})\Delta\mu_{AM} \tag{46}$$

Differentiation of Eq. (46) with respect to $x_{HM}$ results in, with the help of the Gibbs-Duhem relationship,

$$\frac{d\Delta G_{mix}}{dx_{HM}} = \Delta\mu_{HM} - \Delta\mu_{AM} \tag{47}$$

Then it is obtained from Eqs. (43) and (45)

$$\frac{d\Delta G_{mix}}{dx_{HM}} = RT \ln K_{th} - RT \ln K' + RT \ln \frac{x_{HM}}{x_{AM}} \qquad (48)$$

According to the cell model of HM and AM system of cation exchange membranes, the entropy of mixing $\Delta S_{mix}$ and the enthalpy of mixing $\Delta H_{mix}$ are given as follows [73]:

$$\Delta S_{mix} = -R(x_{HM} \ln x_{HM} + x_{AM} \ln x_{AM}) \qquad (49.1)$$

$$\Delta H_{mix} = bx_{HM}x_{AM} \qquad (49.2)$$

where $b$ is constant expressing the energy change of interaction between cation pairs through the mixing process. Hence

$$\frac{d\Delta H_{mix}}{dx_{HM}} = RT(\ln K_{th} - \ln K')$$
$$= b(1 - 2x_{HM}) \qquad (50)$$

Equation (50) indicates that if the mixture of HM and AM can be described by a regular mixture model, then the plot of $\ln K'$ against $x_{HM}$ should become straight line [43]. Figure 41 depicts the plots for H/Li, H/Na, H/K, H/Rb and H/Cs-form membrane systems [71]. A good linearity indicates that $H^+$ and other cations behave as if they are sitting on the cation exchange sites like a regular mixture. This indicates very little interactions between cations in binary cation form Nafion® membranes, i.e., cations do not distinguish each other whether one or the other cationic species comes into the next cation exchange site. The shielding effect of water molecules and sulfonic acid groups surrounding ionic species may account for this behavior.

From the slopes and intercepts in figure 41, $K_{th}$ and $b$ are calculated and summarized for various cation pairs in table 4. $K_{th}$ is a measure of the interaction between cations and sulfonic acid groups, and $b$ is a measure of the interaction of $H^+$ and alkali metal cations $A^+$ in the mixed state as compared with the state before mixing. Larger $K_{th}$ means that $H^+$ has higher affinity to sulfonic acid group than $A^+$ does, which conforms well to the results in figure 40. Also more positive value of $b$ means repulsive (energetically unfavorable) interactions between $H^+$ and $A^+$ as compared with the segregated state of $H^+$-$H^+$ pairs and $A^+$-$A^+$ pairs.

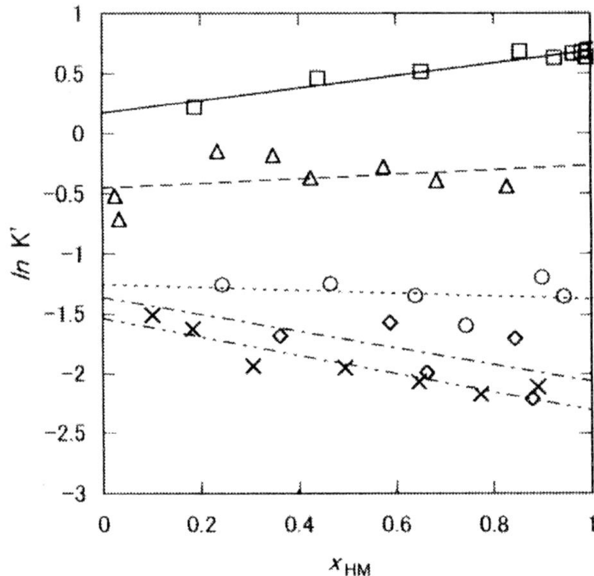



Figure 41. Relations between *ln K'* and the membrane cationic composition $x_{HM}$. Symbols are the same as in Fig. 39 (with permission from the American Chemical Society).

## Table 4. Cation interactions in Nafion® membranes

| Systems | $K_{ex}$ [a] | $K_{th}$ | $b$/J mol⁻¹ |
|---|---|---|---|
| H/Li | 1.85 | 1.54 | 641 |
| H/Na | 0.73 | 0.70 | 235 |
| H/K | 0.25 | 0.27 | -151 |
| H/Rb | 0.22 | 0.18 | -873 |
| H/Cs | 0.16 | 0.15 | -954 |
| Li/Na | 0.46 | 0.52 | -202 |
| Li/K | 0.15 | 0.17 | -175 |
| Li/Rb | 0.17 | 0.19 | -768 |
| Li/Cs | 0.11 | 0.095 | -288 |

$K_{ex}$: Equilibrium constant of the exchange reaction, $K_{th}$: thermodynamic equilibrium constant, $b$: constant expressing the interaction between cations.

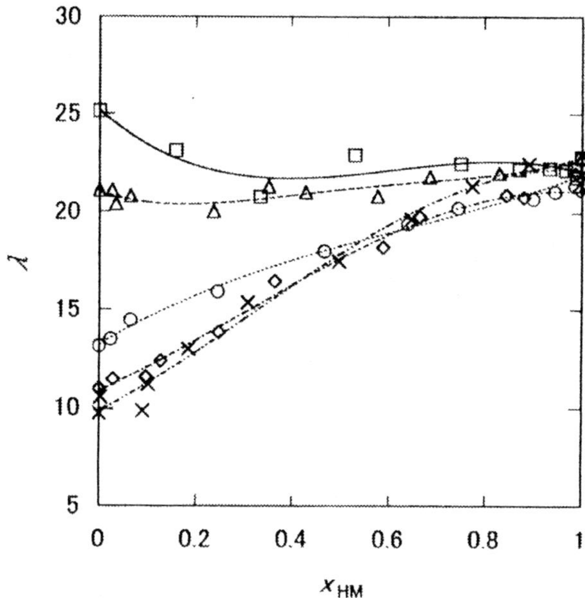

Figure 42. Membrane water content $\lambda$ plotted against membrane cationic composition $x_{HM}$. Symbols are the same as in Fig. 39 (with permission from the American Chemical Society).

It is seen in table 4 that the systems can be classified into two groups, one is for the group of H/Li and H/Na where $b$ is positive (Group 1), and the other is for the group of H/K, H/Rb and H/Cs where $b$ is negative (Group 2). Two kinds of classes are observed also in water content $\lambda$ (figure 42), membrane density in wet state $d_{wet}$, water transference coefficient $t_{H_2O}$ (figure 39) and water permeability $L_p$ expressed as a function of $x_{HM}$ [71]. These differences may be explained by different states of water molecules surrounding these cations.

The water molecules surrounding cations may play an important role in the case of binary cation systems. First the amount of water in the membrane is noted for different cation forms. For Group 1, the amount of water is large, and the repulsive force between a pair of alkali metal cations $A^+$ - $A^+$ would be mitigated by a shielding effect of water sheath. Here the interaction between $H^+$ and $A^+$ cations will become repulsive due to smaller mutual distance, and two kinds of cations avoid mixing each other. On the other hand, for Group 2, water content is low and the polymer structure tends to shrink. In this case the repulsive force between large ions is large, and since the repulsive force between $H^+$ and $A^+$ becomes smaller than for the pairs of same cations, $H^+$ and $A^+$ would mix each other. It is interesting to note that $Li^+$ and $Na^+$ are "structure forming cations" while $K^+$, $Cs^+$ and $Rb^+$ are "structure breaking cations" in aqueous solutions [74], the latter disorients the first hydration sheath and tends to expel water molecules in the ionic channel.

The way the water content changes by the change in membrane composition, as seen in figure 42, might also influence the amount of dragged water in mixed cation systems (figure 39). In Group 2, the water content $\lambda$ decreases in the mixed state, while in Group 1, $\lambda$ stays almost constant. In Group 2, some of the water molecules are discarded when the mixture of ions drags water, while in Group 1 two kinds of cations drag fixed amount of water molecules (figure 39). This may also be considered as specific characteristics of the microscopic structure of PFSI, where water and cations interact in a way to maintain the high ionic conductivity through common pathways of ionic channels.

## 5.9. ALCOHOL PENETRATION AND TRANSPORT PROPERTIES

Direct methanol fuel cells (DMFCs) have received increasing attraction as various power sources for portable devices such as cellular phones, personal digital assistants (PDAs), laptop computers, etc. [75,76]. So far many studies have been devoted to the development of DMFCs with higher performances. However, methanol crossover through electrolyte membranes such as Nafion® and others is one of the critical problems in practical use because of the fuel loss, lowering of the cell voltage and power density due to deterioration of oxygen reduction reaction at the cathode, etc [76]. To design advanced electrolyte membranes for DMFC, investigations on the methanol diffusion behaviors and influences of methanol on proton transport in the PFSI membranes are quite valuable. In this section, methanol permeability $P_M$ and its influence on the proton conductivity $\kappa$ will be discussed using H-form PFSI membranes with different $EW$ values.

Figure 43 shows the $P_M$ values for membranes of various $EW$ values [77]. The $P_M$ value increases with decrease of the $EW$ value for all the membranes. This implies that methanol molecules penetrate into the hydrophilic regions to form ionic cluster regions together with water molecules and sulfonic acid groups, in a similar way to the fully hydrated membranes, and diffuse through the expanded ion cluster domains. It appears that $P_M$ is strongly polymer structure dependent aside from $EW$ dependence. The difference in the length of the main and/or side chains of polymers may cause to modify the size and distribution of ion cluster regions in the membrane, and induce the changes in the transport rate of methanol.

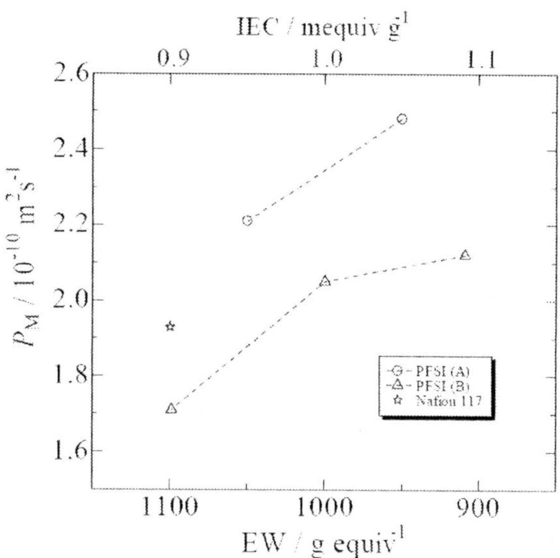

Figure 43. Methanol permeability $P_M$ for the 6 different membranes. PFSI (B)-1000-2 was used for the measurements on PFSI (B)-1000 (with permission from Elsevier Science).

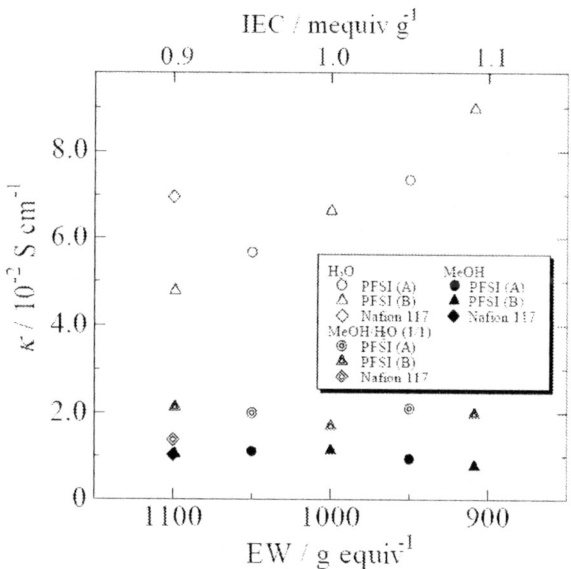

Figure 44. Proton conductivity $\kappa$ for the 18 different membranes in the $CH_3OH$, $CH_3OH/H_2O$, and $H_2O$ penetrated states. PFSI (B)-1000-2 was used for the measurements on PFSI (B)-1000 (with permission from Elsevier Science).

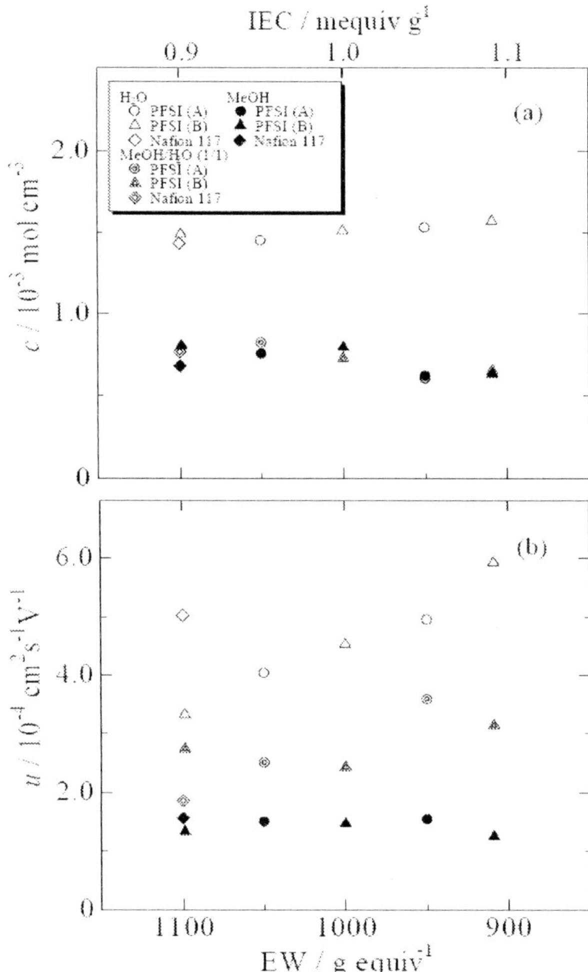


Figure 45. The concentration $c$ (a) and mobility $u$ (b) of carrier protons in the 18 different membranes of $CH_3OH$, $CH_3OH$ /$H_2O$, and $H_2O$ penetrated states. PFSI (B)-1000-2 was used for the measurements on PFSI (B)-1000 (with permission from Elsevier Science).

Proton conductivity $\kappa$ becomes lower when methanol penetrates the membrane especially in smaller $EW$ membranes (figure 44), i.e., presence of methanol causes not only the fuel loss but also the decrease in the $\kappa$ value. It should be noted that the methanol/water mixture-penetrated membranes show similar behavior to the methanol-penetrated ones. This is a quite serious problem for the DMFCs because even small amount of methanol causes large decrease in

$\kappa$. Figure 45 (a) and (b) show the concentration ($C_{H^+}$) and the mobility ($u_{H^+}$) of proton, respectively, for different membranes. By methanol penetration, $C_{H^+}$ becomes about a half of the hydrated membranes, and $u_{H^+}$ also decreases especially for lower $EW$ membranes. It is observed that $u_{H^+}$, rather than $C_{H^+}$, is the major factor that determines the proton conductivity.

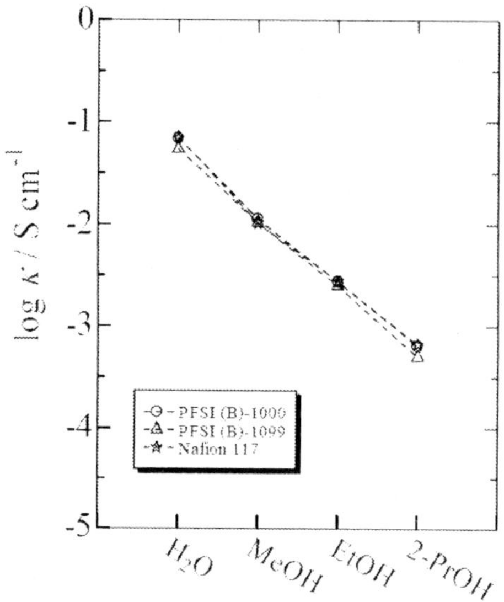


Figure 46. Proton conductivity $\kappa$ for 3 different membranes in fully water- or alcohol-penetrated states (with permission from the American Chemical Society).

The difference of the molecular structures between methanol and water is that CH$_3$ group substitutes H. For the purpose of investigating the influence of the alkyl group on proton transport in the membranes, the proton conductivity for membranes penetrated with water and three kinds of alcohols is depicted in figure 46. The proton conductivity decreases seriously as alcohols substituted with large alkyl group are penetrated. Therefore, the larger alkyl groups are one of the factors preventing the faster proton transport.

Figure 47 (a) shows the $D_{CH_3}$ and $D_{OH}$ for the alcohol-penetrated membranes against the $EW$ value measured by PGSE-NMR [78]. Here $D_{CH_3}$ corresponds to the self-diffusion coefficient of methanol and $D_{OH}$ represents the weighted average of self-diffusion coefficients of alcoholic OH and the H$^+$

($H_3O^+$). The $D_{CH_3}$ values are around $10^{-10}$ $m^2s^{-1}$, which is in similar order as in the methanol penetrated Nafion$^®$ 117 reported by Ren et al. [65]. The $D_{OH}$ is found to be always larger than the $D_{CH_3}$, and both values increased with decreasing the $EW$ value. In addition, the difference between $D_{CH_3}$ and $D_{OH}$ shown in figure 47 (b) also increases with decreasing the $EW$ value, especially for membranes penetrated by the smaller alcohols. The alcohols can transport only by the vehicle mechanism [34]. Since a proton can transport through the OH network of neighboring alcohols, the proton transport by the Grotthuss (hopping) mechanism [67] will be favorable for the smaller alcohol. On the other hand, for the membranes penetrated by larger alcohols, the vehicle mechanism will become more significant. Therefore, the difference between $D_{CH_3}$ and $D_{OH}$ progressively diminishes for the larger alcohol. The interaction energies between proton and alcohols as calculated by density functional theory support this mechanism [78].

The temperature dependence of diffusion coefficients of the alcohol-penetrated membranes is studied by PGSE-NMR in the temperature range of 30 to -40°C [78]. Figure 48 shows the Arrhenius plots of the $D_{CH_3}$ and $D_{OH}$ for (a) $CH_3OH$, (b) $C_2H_5OH$ and (c) $2\text{-}C_3H_7OH$ in the membranes along with pure liquids. The $D_{CH_3}$ and $D_{OH}$ in the pure liquids almost coincide each other, but in the membranes discrepancies are observed as discussed above. All the $D_{CH_3}$ in the membranes are one order of magnitude smaller than the values in pure liquids, indicating that the ion cluster structures suppress the diffusion rate of alcohols in the membrane. $D_{OH} > D_{CH_3}$ holds for all the membranes, and the smaller $EW$ membranes show faster diffusion for alkyl and OH (including $H^+$) species. The order of the diffusion coefficients is $CH_3OH > C_2H_5OH > 2\text{-}C_3H_7OH$ at all temperatures between 30 and -40 °C.

The activation energies $E_a(CH_3)$ and $E_a(OH)$ are calculated from the Arrhenius plots in the membrane and in pure liquids (figure 49). Comparing with the bulk alcohols, all the alcohols in the membranes show larger $E_a(CH_3)$ values, which increases with increasing the alcohol size. The $E_a(OH)$ are slightly smaller than $E_a(CH_3)$ for all the membranes. On the other hand, the activation energies show only small $EW$ dependence.

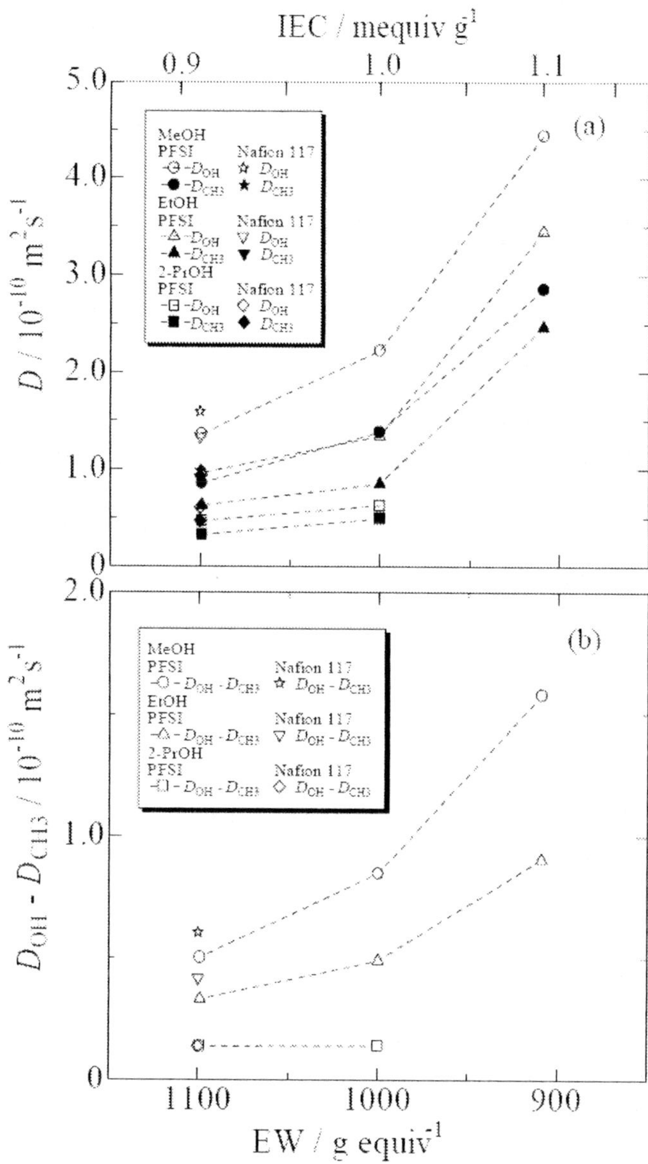



Figure 47. EW dependence of (a) the self-diffusion coefficients $D_{OH}$ and $D_{CH_3}$, and (b) the difference $D_{dif}$ between the $D_{OH}$ and $D_{CH_3}$ for the 4 different membranes measured at 30 °C. PFSI (B)-1000-2 was used for the measurements on PFSI (B)-1000 (with permission from the American Chemical Society).

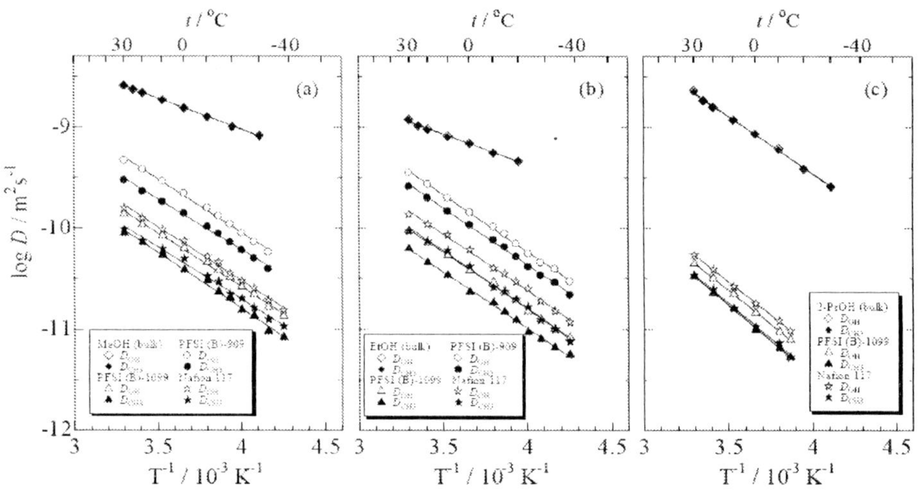

Figure 48. The Arrhenius plots of self-diffusion coefficients $D_{OH}$ and $D_{CH_3}$ for the alcohol-penetrated membranes having various EW values and the pure alcohols for (a) $CH_3OH$, (b) $C_2H_5OH$ and (c) $2\text{-}C_3H_7OH$ (with permission from the American Chemical Society).

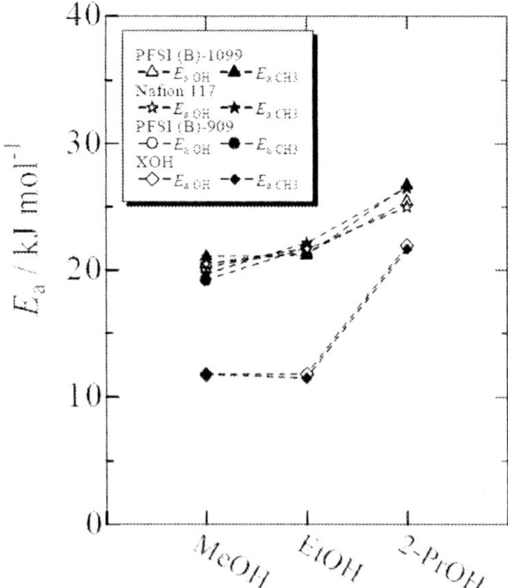

Figure 49. The activation energy $E_a$ from the slope of Arrhenius plots of $D_{CH_3}$ and $D_{OH}$ in the alcohol-penetrated membranes (with permission from the American Chemical Society).

In the PESI membranes, the alcohol diffusion rate and the proton mobility are in a trade-off relationship, and the membrane having higher proton conductivity shows higher alcohols crossover. This is due to the ionic cluster regions formed in the membranes. Because alcohols move in the expanded space by the vehicle mechanism, the membrane swelling is a crucial factor for diffusion of species. Since proton can move relatively faster than alcohols by Grotthuss mechanism, membranes that show lower swelling might be a solution to the alcohol crossover problems. The design and control of the morphology is quite important, for example, by the crosslinking the polymer chain to improve the properties. In addition, the alcohol with lower molecular weight should be used for higher proton mobility in the membranes.

# Chapter 6

# DECISIVE FACTORS IN PROTON CONDUCTION AND THE MEMBRANE DESIGN AND FUEL CELL CONTROL

## 6.1. CORRELATIONS OF PARAMETERS

Table 5 summarizes cationic mobility and relevant parameters obtained in this study. Three major factors will be discussed that determine the cation mobility in the membrane:

1. larger affinity of cations to the sulfonic acid groups may cause ion-pair formation, and result in smaller cationic mobility,
2. smaller water content in the membrane may lead to smaller ionic channels, and result in smaller cationic mobility, and
3. larger water transference coefficient $t_{H_2O}$ may cause a higher friction forces against cation movement, and result in smaller cationic mobility.

(1) is related to the degree of electrostatic interaction, and (2) and (3) are related to the relative size of moving ions and the ionic channel.

In figure 50 (a), cation mobilities are plotted against $K_{th}$, the interaction parameter between the cation and sulfonic acid groups, obtained from H/A mixed form membranes (A = alkali metal cations). A good correlation is observed in the figure, which means that the membrane conductivity is determined first by the electrostatic force between cations and sulfonic acid groups. Li/A mixed cation systems show the same tendency. Cation-cation interactions are not critical in the membranes because of the shielding effect [71,72].

**Table 5. Cationic mobility and relevant parameters in Nafion® membranes (with permission from the American Chemical Society)**

| Ion $M^+$ | $\lambda$ in M-form | $t_{H_2O}$ in M-form | $K_{th}$ for $H^+/M^+$ | $u_i$ $10^{-8}$ m$^2$V$^{-1}$s$^{-1}$ | $u_i(l)^a$ $10^{-8}$ m$^2$V$^{-1}$s$^{-1}$ |
|---|---|---|---|---|---|
| $H^+$ | 22.0 | 2.6 | 1.00 | 19.4 | 36.3 |
| $Li^+$ | 25.1 | 15.8 | 1.54 | 3.51 | 4.01 |
| $Na^+$ | 21.3 | 9.2 | 0.70 | 4.50 | 5.19 |
| $K^+$ | 13.1 | 4.9 | 0.27 | 3.45 | 7.62 |
| $Rb^+$ | 10.9 | 5.5 | 0.18 | 2.59 | 7.92 |
| $Cs^+$ | 9.7 | 5.2 | 0.15 | 2.20 | 7.96 |

$^a$ Ionic mobility in the liquid electrolyte at infinite dilution.

Note the concentration of the sulfonic acid groups $c_{SO_3^-}$ per volume of the membrane. When anionic site number is increased (*EW* is decreased), the membrane conductivity improves. In this case $c_{SO_3^-}$ stays almost unchanged because the water content in the membrane also increases. The membrane conductivity increases not due to the increase of $c_{SO_3^-}$ but due largely to the increase of the proton mobility $u_{H^+}$ [66], by altered domain structure of ion clusters [23].

Figure 50 (b) shows the relationship between the cation mobility $u_i$ and the parameter $\lambda - t_{H_2O}$ obtained for H/A and Li/A cation form membranes [71,72]. A good correlation means that cation mobility is determined by the amount of water residing in the ionic channel. The role of water molecules in ionic channels may be emphasized here. The larger the domain of ion conducting path, the higher will be the cationic mobility [57]. Thus the volume fraction of water may be an important parameter in the membrane [71]. The binary cation systems provide good discussions on this point. If the second cation is less hydrophilic than the host cation, this will bring about the smaller volume of hydrophilic domains, and the membrane shrinks and density increases. It is observed that conductivity declines more when less hydrophilic cations enters into the membrane [72].

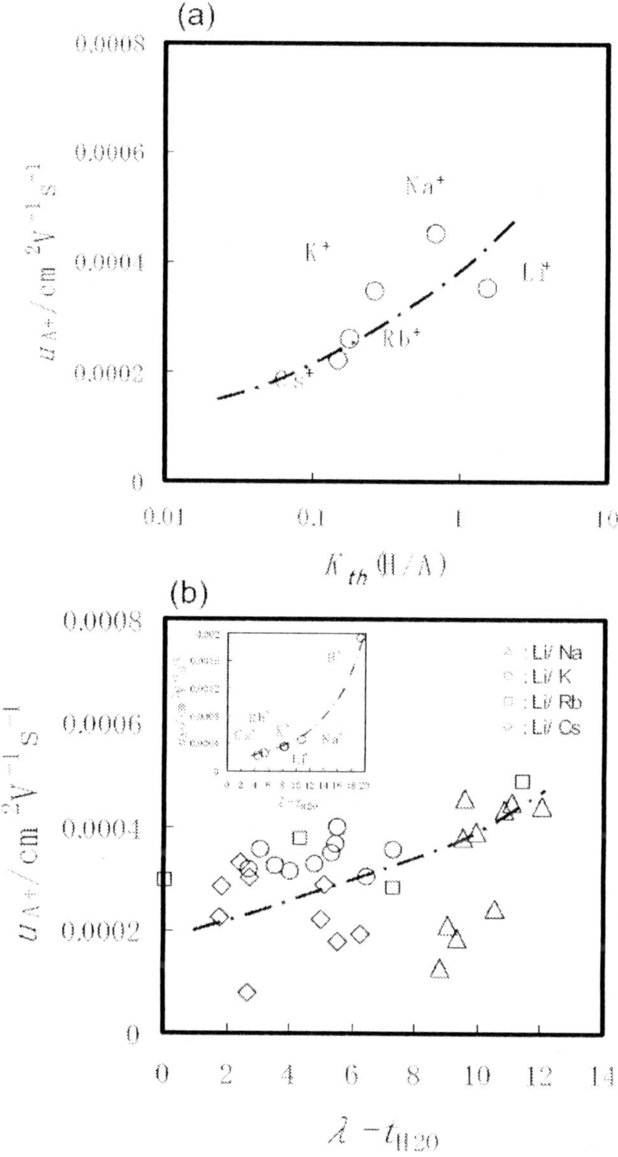



Figure 50. Cation mobilities $u_{A^+}$ obtained for various cation mixtures. (a) $u_{A^+}$ plotted as a function of $K_{th}$. (b) $u_{A^+}$ plotted as a function of the parameter $\lambda - t_{H_2O}$. The inset plots are from single cation form systems (with permission from Elsevier Science).

In the narrow ionic channels, cations that have high $t_{H_2O}$ value would suffer a large friction force from the surroundings, and conductivity would become low. The irregularity of $Li^+$ ion in table 5, i.e., the order of mobility in the membrane, $Na^+ > Li^+ > K^+ > Rb^+ > Cs^+$, can thus be explained, if the large $t_{H_2O}$ value of $Li^+$ is taken into account.

It appears that the friction force acting on cations together with electrostatic force between cations and sulfonic acid groups determines the cation mobility. These factors should be significant criteria in designing micro-structures of high performance polymer membranes, but a more essential factor would be the structural features of ionic channels in PFSI that shaped under the phase separation processes of polymer electrolytes.

## 6.2. DESIGN CONCEPTS FOR HIGH PERFORMANCE MEMBRANES

Designing new concept membranes of high performances is a big current issue in PEFC technologies. Although Nafion® and other types of PFSI membranes are standard materials for fuel cell applications, some problems, especially high cost and complicated fabrication process, are still drawbacks for commercialization. For finding alternatives to PFSI membranes, new strategies for polymer synthesis are presented [79]. In this section, major directions of future designing of high ion conductivity polymer electrolytes for PEFC and other applications will be attempted, to support such efforts. Important specifications derived from the discussions in preceding sections are noted here:

1.  a high density of sulfonic acid groups to increase the ionic site density in the hydrophilic domains,
2.  a high water content to enlarge hydrophilic domains in the membrane,
3.  alignment of channel structures to ease ion (and water) transport.

Both the points (1) and (2) would include a risk of lowering the mechanical properties of the membrane, so it is important to seek for the optimization when one designs the polymer structure. High density of ion exchange groups often induces large membrane swelling, which will lower the thermal and chemical durability. This is one of the major problems for hydrocarbon membranes developed as alternatives. Taking points (1) and (2) at the same time is in fact a difficult task, without sacrificing the membrane durability.

To attain a high ionic conductivity, channel structure of ion conducting path in PFSI may be a model to be followed. This direction can be inferred from table 1, which shows a comparison between Nafion® 117 and crosslinked cation exchange membrane CR61 AZL 386. In spite of higher ion exchange capacity, CR61 AZL 386 shows lower conductivity than Nafion® and this may be ascribed to the phase separated polymer structures of PFSI as discussed in section 6.1. Tailoring polymer structures that adhere to the concept (3) may be pursued by taking a compromise between channel and crosslinking structures, which will be discussed further in Section 3 (HYDROCARBON POLYMER ELECTROLYTES FOR FUEL CELL APPLICATIONS).

At the present it is interesting to refer to suggestions made by Kreuer et al. in designing low-cost alternatives for PFSI membranes in PEFC applications [79]. It is proposed that the ideal polymer microstructure is described by following points: wide ionic channels, more separated phase structure, less branched cluster domain, good connectivity (less dead-end channels), small $-SO_3^-/-SO_3^-$ separation and $pK_a$ value of around -6. According to this criteria, the wide channel and connectivity would be important keywords, but the high water content will not be a necessary point, especially for future membranes for middle temperature PEFC operations (between 150 and 200 °C).

In designing membranes for DMFC applications, the problem of methanol crossover should be resolved in addition to the above-mentioned points. Generally, hydrocarbon ionomer membranes are considered to be better candidates for the DMFCs than PFSI membranes because of significant swelling by methanol of the latter. Some of hydrocarbon membranes show high proton conductivities comparable to PFSI membranes. Due to the absence of clear ionic cluster regions, they show lower methanol crossover properties. The advantages of the hydrocarbon membranes are lower cost and easier synthesis processes. It should be noted that a proton moves more easily than methanol, in the narrow ion conducting paths in membranes. The design and control of the morphology of the ionomer chains is quite important, for example, by the crosslinking to improve the properties.

## 6.3. EFFECTS OF IMPURITY IONS ON PEFC PERFORMANCES AND ITS DEGRADATION

When any kind of cations enter into the MEA as foreign impurity substances, these cations will preferentially be exchanged with inherent $H^+$ in the polymer electrolyte membranes in PEFC, as suggested in figure 40. Impurities may come into the fuel cell chamber either as very fine salt nanoparticles from the atmosphere, or as corrosion products from tubing or stack materials. As direct impact of impurity cations on the membrane performances in PEFC, the following three factors should be considered in membranes.

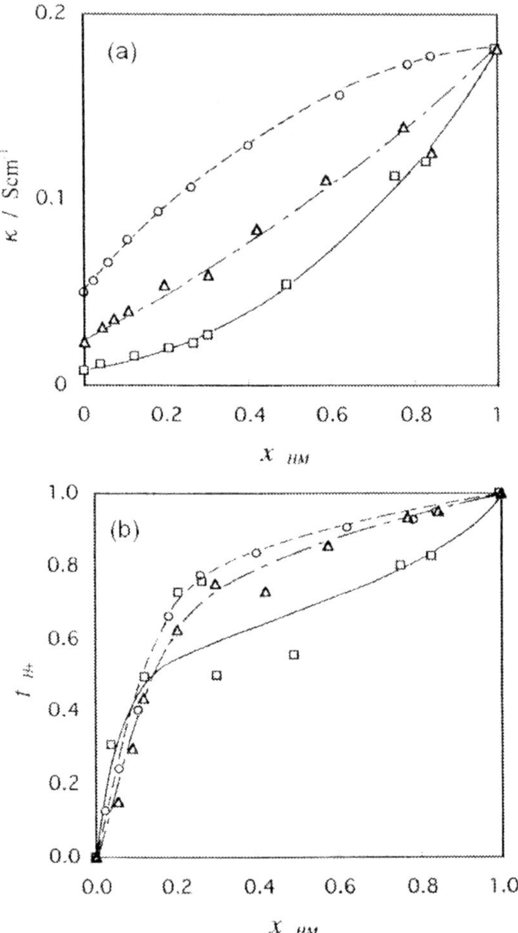

Figure 51. (a) Membrane conductivity $\kappa$ against cationic composition $x_{HM}$ in Nafion® 115 membranes. (b) Ionic transference number of H⁺ $t_{H^+}$ against cationic composition $x_{HM}$ in Nafion® 117 membranes. (□), H/Fe system; (△), H/Ni system; (○), H/Cu system (with permission from the American Chemical Society).

One is decrease in ionic conductivity which induces the increase of Ohmic resistance polarization in the fuel cell. Figure 51 (a) shows such cases, and membranes contaminated with $Fe^{3+}$, $Ni^{2+}$ and $Cu^{2+}$ ions cause serious conductivity decrease. Especially $Fe^{3+}$ has very high tendency to enter into the membrane (figure 40), and possibly will cause most serious degradation in ionic conductivity. Since $Fe^{3+}$ is known to catalyze the production of OH radicals from $H_2O_2$ generated in the anode or the cathode, careful design of PEFC is required to get rid of this ion during the fuel cell operation.

Second is decrease of $H^+$ transference numbers in the membrane. The example is depicted in figure 51 (b), which shows that $t_{H^+}$ gradually decreases when impurity cations enter into the membrane. Decrease of $t_{H^+}$ causes not only the concentration polarization due to depletion of $H^+$ at the cathode but also the membrane diffusion potential difference between the anode and the cathode [80].

The third is depletion of water content $\lambda$ by the change of inherent ionic cluster structures in the polymer (see figure 42). Decrease of $\lambda$ causes decrease of membrane ionic conductivity, as shown in figure 17, resulting in the Ohmic resistance polarization. All these phenomena are consequences of static membrane properties by impurity ions, and the impact is relatively small as compared with dynamic degradation phenomena, which occur as a result of altered flux of water. Details of such phenomena are described in the next section.

## 6.4. IMPLICATIONS FOR OPTIMUM CONTROL OF PEFC: APPLICATION TO WATER MANAGEMENT

In PEFC, maintaining high power and efficient operation are a major task for commercial running. From the viewpoint of the membrane performance, this means to maintain the membrane resistance overpotential minimal, by keeping the membrane in well-hydrated state, because the resistance of the membrane depends strongly on the water content in the membrane (see figure 17). This control problem is called "water management" in PEFC operations, and contains avoiding both membrane drying and water flooding at the anode and the cathode catalyst sides, respectively. Poor control would cause high Ohmic resistance or serious gas diffusion limitations in the fuel cell.

Simulations of PEFC have been handled so far mainly considering transport equations in the gas flow channel and gas diffusion layers, polarization at the catalyst layers, etc. However, due to very complicated sets of equations used in calculations, only numerical evaluations have been performed for the purpose of predicting the cell performances with fixed operation parameters [81]. In this section we approach the problem by paying attention to the phenomena in the membrane electrolyte layer. First we solve the transport equations analytically and discuss water management by changing parameters in a systematic way. This approach would be useful for the diagnosis of water management, which is made through the comparison of relative significance of parameters.

The water flux in the membrane is expressed by the following equation, based on the "solution diffusion model":

$$j_{H_2O} = -D_{H_2O}\frac{\partial C_{H_2O}}{\partial x} + t_{H_2O}\frac{j}{F}$$ (51)

here $c_{H_2O}$ is water concentration inside the membrane, and $j$ is current density. At high current operations, the membrane tends to dry from the anode side by electro-osmotic drag of water while water generated at the cathode counteracts it by back-diffusion. Solving Eq. (51) with appropriate boundary conditions gives the water profile inside the membrane, with which the hydration state is controlled from outside.

Consider a polymer electrolyte membrane in PEFC, which is bounded by two catalyst layers, an anode and a cathode, at each side. The interface between the anode and the membrane is set as the $y$-$z$ plane, and $x$-axis extends from the anode to the cathode into the membrane. It is assumed that the membrane is homogeneous, and therefore uniform in the $y$-$z$ direction. The one-dimensional linear flux equation of water in the bulk of the polymer electrolyte is written in the form [82]:

$$\frac{\partial c_{H_2O}}{\partial t} = -\frac{\partial j_{H_2O}}{\partial x} = \frac{\partial}{\partial x}(D_{H_2O}\frac{\partial c_{H_2O}}{\partial x} - \frac{j}{F}t_{H_2O})$$ (52)

This gives at the steady state

$$D^{(0)}\frac{\partial^2 c_{H_2O}}{\partial x^2} - \frac{jt^{(1)}}{F}\frac{\partial c_{H_2O}}{\partial x} = 0$$ (53)

where $D^{(0)}$ and $t^{(1)}$ are the factors of water concentration dependence in the parameters:

$$D_{H_2O} = D^{(0)} + D^{(1)}c_{H_2O} + D^{(2)}c_{H_2O}^2 + ... \tag{54.1}$$

$$t_{H_2O} = t^{(0)} + t^{(1)}c_{H_2O} + t^{(2)}c_{H_2O}^2 + ... \tag{54.2}$$

Then the water concentration in the membrane for the steady state is expressed by [83]

$$c_{H_2O}(x) = c_0 - A - B\exp\left(\frac{It^{(1)}}{D^{(0)}F}x\right) \tag{55}$$

Constants $A$ and $B$ can be determined from boundary conditions [83]. The continuity of water flux at the anode│membrane interface requires:

$$\frac{v_1 j}{F} + k_1\{c_1 - c_{H_2O}(0)\} = -D^{(0)}\frac{\partial c_{H_2O}(0)}{\partial x} + \frac{jt_1^{(0)}}{F} + \frac{jt_1^{(1)}}{F}c_{H_2O}(0) \tag{56}$$

here $v_1$ is the factor expressing the rate of $H_2O$ entry proportional to the current density $j$, and $k_1$ is the mass transfer coefficient for the $H_2O$ entry in accordance with the Henry's law penetration. The subscript 1 refers to the parameters at the anode│membrane interface.

For the boundary condition at the cathode│membrane interface, two kinds of cases are considered:

(B.C.1) Water concentration is constant at the cathode which locates at distance d from the anode:

$$c_{H_2O}(d,t) = c_0 \tag{57}$$

(B.C.2) Water penetration is allowed at the cathode which locates at distance d from the anode:

$$\frac{v_2 j}{F} + k_2\{c_2 - c_{H_2O}(d)\} = D^{(0)}\frac{\partial c_{H_2O}(d)}{\partial x} - \frac{jt_1^{(0)}}{F} - \frac{jt_1^{(1)}}{F}c_{H_2O}(d) \tag{58}$$

where the subscript 2 refers to the parameters at the membrane│cathode interface.

Some base-case parameters used for calculation are listed in table 6 based on measured values for Nafion® membranes. Simulated results for the net water flux $\phi_{H_2O} \equiv j_{H_2O} / j$ and the membrane resistance overpotential $\eta$(mem) are tested with experimental results. In spite of very rough estimation in the present analysis, reasonable agreement is obtained with a single adjustable parameter $k$ [83]. Predicting the membrane hydration state is normally a very difficult task, but it is handled theoretically and visualized easily, if the transport parameters of ions and water in the membrane are known.

**Table 6. Specific parameters of the Nafion® membrane used in the calculation (with permission from Elsevier Science)**

| Parameters | Notation/unit | Method | Value at 80 °C | Refs. |
|---|---|---|---|---|
| Density, dry state | $d$(dry)/g cm$^{-3}$ | Gravimetric | 2.02 (25 °C) | 37 |
| Volume ratio, wet vs. dry | $\chi_V$ | Measuring size | 1.62(25 °C) | 37 |
| Equivalent weigh | E.W./g mol$^{-1}$ | Titration | 1100(25 °C) | |
| Water content at 100 % water vapor | $\lambda = n(H_2O)/n(SO_3^-)$ | Gravimetric | 14(25 °C) | 9 |
| Water concentration | $c_{H2O}$/mol cm$^{-3}$ | $c_{H2O} = d$(dry) $\lambda$/ $EW\ \chi_V$ | $1.13\times10^{-3}$ (25 °C) | 83 |
| Diffusion coefficient of water, H-form at $\lambda$=14 | $D_{H2O}$/cm$^2$ s$^{-1}$ | Streaming potential | $1.25\times10^{-5}$ | 9 |
| Diffusion coefficient of water, Na-form at $\lambda$=14 | $D_{H2O}$/cm$^2$ s$^{-1}$ | Streaming potential | $5.0\times10^{-6}$ | 9, 83 |
| Water transference coefficient, H-form | $t_{H_2O,1}$ | Streaming potential | 3.2 | 83 |
| Water transference coefficient, Na-form | $t_{H_2O,2}$ | Streaming potential | 10.2 (60 °C) | 83 |
| Specific conductivity, H-form | $\kappa_1$/S cm$^{-1}$ | AC impedance | 0.33 (50 °C) | 83 |
| Specific conductivity, Na-form | $\kappa_2$/S cm$^{-1}$ | AC impedance | 0.099 (50 °C) | 83 |

The dependence of the water concentration at the anode | membrane interface $c(0)/c_0$, net water flux $\phi_{H_2O}$ and the membrane resistance overpotential $\eta$(mem) on current densities at various values of the membrane thickness $d$ is shown in figure 52. The thickness of the membrane affects largely the membrane performance, because of the depleted water content by changed water fluxes for thick membranes. Thinner membranes are preferable in this context, especially less than 60 μm, if not for the increased gas crossover between the anode and the cathode. The advantage of the present methodology compared to the sophisticated numerical analyses using more realistic two-dimensional or three-dimensional models is its simple formula. The explicit analytical solutions are easy to use in systematic evaluation of transport related parameters for water management diagnosis.

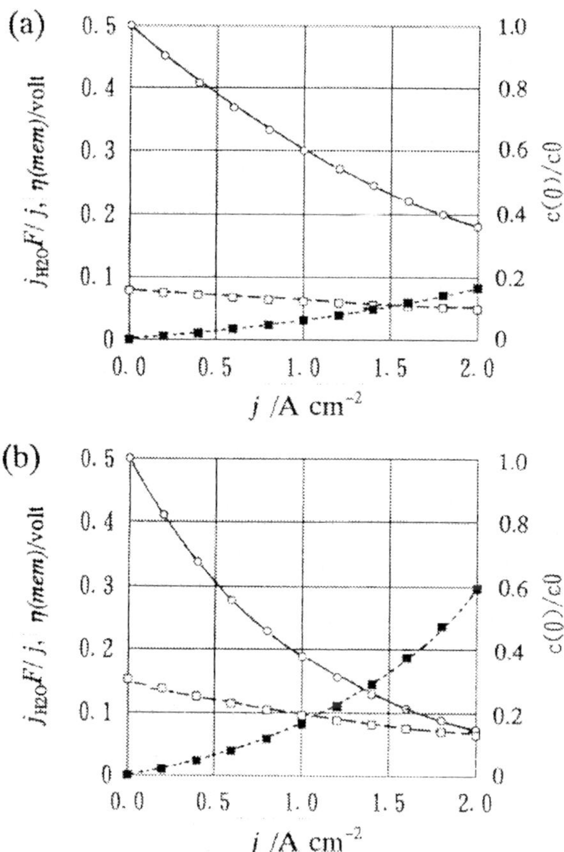

Figure 52. The effect of current density on the water concentration at the anode│membrane interface $c(0)/c_0$ (◯), net water flux $\phi_{H_2O}\left(\equiv j_{H_2O}/j\right)$ (□) and the membrane resistance overpotential $\eta(mem)$ (■) at the membrane thickness $d = 50$ μm (a) and 100 μm (b) (with permission form Elsevier Science).

The problem about the water management becomes significant when some impurity cations enter into the membrane [82]. The experimental fact shows that, when H-form membrane is contacting foreign impurity cations, $D_{H_2O}$ decreases while $t_{H_2O}$ increases [37]. It is then inferred that osmotic drag of water increases while back diffusion decreases in Eq. (51), and water starts to deplete from the anode side of the membrane. In order to solve the problem of impurity effects on membrane drying, "infected zone" is assumed as specific boundary conditions, and water transport equation is solved analytically to simulate the water flux and water content profiles in the membrane [84,85].

First the case where the impurity ions $Q^{n+}$ penetrate into the anode side of the membrane is considered (figure 53 (a)). Suppose that the "infected zone" extends from the anode ($x = 0$) to thickness $\delta$ in the membrane, and beyond $\delta$ to the cathode ($x = d$), the membrane is in pure H-form. The ionic fraction of $Q^{n+}$ in the membrane, $X_{QM}$, is assumed as follows:

$$X_{QM} = q(1 - \frac{x}{\delta}), \; X_{HM} = 1 - q(1 - \frac{x}{\delta}) \qquad (0 < x < \delta) \qquad (59.1)$$

$$X_{QM} = 0, \; X_{HM} = 1 \qquad (\delta < x < d) \qquad (59.2)$$

Transport equations at each portion of the membrane are described by

$$\frac{\partial}{\partial x}\left(\left[D_2^{(0)} + \left(D_1^{(0)} - D_2^{(0)}\right)\left\{1 - q(1 - \frac{x}{\delta})\right\}\right]\frac{\partial c_{H2O}}{\partial x}\right)$$

$$-\frac{j}{F}\frac{\partial}{\partial x}\left(\frac{1}{\frac{1-q}{q} + p + (1-p)\frac{x}{\delta}}\left[(t_1^{(0)} + t_1^{(1)}c_{H2O})\frac{1-q}{q} + (t_1^{(0)} + t_1^{(1)}c_{H2O})p + \{t_1^{(0)} - t_2^{(0)}p + (t_1^{(0)} - t_2^{(0)}p)c_{H2O}\}\frac{x}{\delta}\right]\right)$$

$$= 0$$
$$(0 < x < \delta) \qquad (60.1)$$

$$D_1^{(0)}\frac{\partial^2 c_{H2O}}{\partial x^2} - \frac{j}{F}t_1^{(1)}\frac{\partial c_{H2O}}{\partial x} = 0 \qquad (\delta < x < d) \qquad (60.2)$$

where $p$ is defined as the ratio of mobilities in the membrane:

$$p \equiv \frac{u_{Q^{n+}}(m)}{u_{H^+}(m)} \tag{60.3}$$

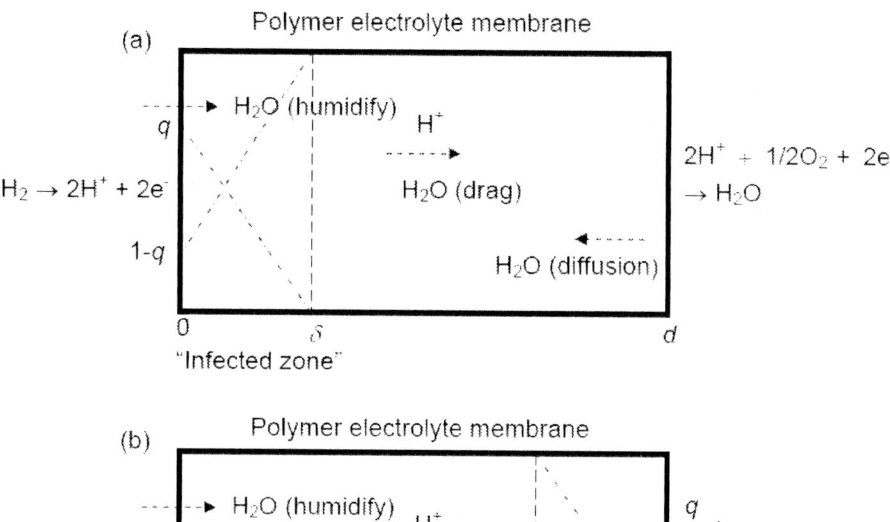


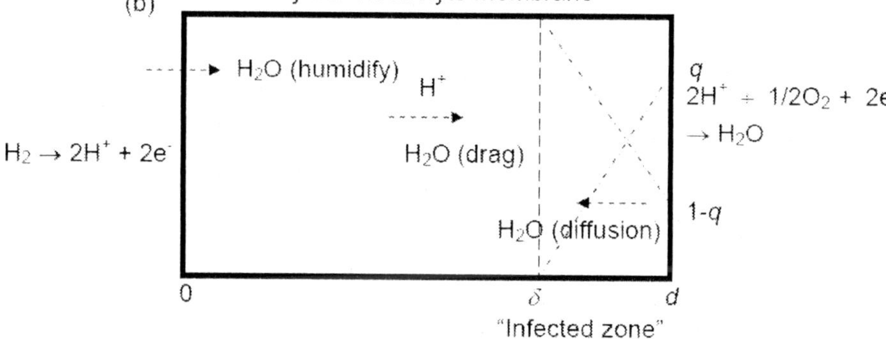


Figure 53. Illustration of concentration profiles of contaminant and $H^+$ ions in polymer electrolyte membranes. (a), Anode side contamination, (b), cathode side contamination (with permission from Elsevier Science).

and the diffusion coefficient of water and the water transference coefficient are assumed to alter by the presence of impurity ions.

$$D_{H2O} = D_q^{(0)} = X_{HM} D_1^{(0)} + H_{QM} D_2^{(0)} \tag{61.1}$$

$$t_{H2O} = t_{H+}(m)t_{H2O}(HM) + t_{Q^{n+}}(m)t_{H2O}(QM) \tag{61.2}$$

where $D_1^{(0)}$ and $D_2^{(0)}$, and $t_{H2O}(HM)$ and $t_{H2O}(QM)$ are related parameters in H-form and Q-form membranes, respectively.

The case where the impurity ions enter into the cathode chamber is treated in the same way. The change in the ionic fraction of $Q^{n+}$ and $H^+$ through the membrane from the anode to the cathode is expressed as follows (figure 53(b)):

$$X_{QM} = 0, X_{HM} = 1 \qquad (0 < x < \delta) \qquad (62.1)$$

$$X_{QM} = q(\frac{x-\delta}{d-\delta}), X_{HM} = 1 - q(\frac{x-\delta}{d-\delta}) \qquad (\delta < x < d) \qquad (62.2)$$

Solving the above equations with appropriate boundary conditions leads to complicated forms of Gauss' hypergeometric functions, but once the formulas are established, analytical calculations are easily performed to know the hydration and conductivity in the membrane [84]. In simple cases the stepwise concentration profiles can be used in the infected zone.

The results are depicted in figure 54 for the cathode contamination, where parameters are systematically changed and displayed for the relative water content, $\lambda/\lambda_0$, net water flux $\phi_{H_2O}(\equiv j_{H_2O}F/j)$ and membrane resistance overpotential $\eta(mem)$. Left-hand side figures show the results for no contamination while right-hand side figures show the results where 10% of ion exchange sites are contaminated by $Na^+$ ( for $\delta = 0.1d$ ). Clearly, the contamination of membrane causes the decrease in the water content, the increase in the net water flux and the increase in the membrane resistance overpotential. This effect becomes apparent when the ion exchange sites are occupied by contaminant ions more than 5%. The effect of each parameter on the membrane performance is discussed in the following.

## 1. Effect of Current Density j (Figure 54-1)

The increase in the current density causes membrane drying and high membrane resistance overpotential, directly due to the water drag by electro-osmosis. The important fact is that this overpotential loss increases non-linearly with the increase in $j$. A trade-off relationship should be taken into account between the cell power and the efficiency loss owing to the membrane drying. The trend becomes even larger as the membrane is contaminated more by foreign ions.

## 2. Effect of Humidification Parameter k (Figure 54-2)

The increase of $k$ accelerates water vapor penetration into the membrane through the anode│membrane interface, which results in the increase in the water content, the increase in net water flux and the decrease in the membrane resistance overpotential. However, in order to suppress the effect of membrane contamination, $k$ needs to be increased more than 10-fold, which is not an efficient task. Membrane contamination should therefore be avoided at any cost in this context.

## 3. Effect of Humidification Parameter v (Figure 54-3)

The effect of $v$ is less apparent as compared to $k$. As can be seen in Eq. (56), increase of $v$ counteracts water drag, but in the range $0 < v < 1$, this counteraction seems to be only small in the sense of suppressing the contamination effect.

## 4. Effect of the Extent of Contamination q (Figure 54-4)

At low current densities the effect of contamination $q$ is not apparent, but at high current density operations it becomes more serious (infected zone at the cathode is assumed to be 10% of the membrane thickness). It appears that membrane resistance increases both by membrane drying and by lowered conductivity due to contamination. When about 70% of ion exchange sites in the infected zone are exchanged with contaminant ions, there is a serious effect on the membrane performance.

## 5. Effect of Membrane Thickness d (Figure 54-5)

As in the case of current density, the membrane performance depends largely on the membrane thickness. It should be noted that the membrane resistance overpotential is not a linear function of membrane thickness, but of higher order dependence because of membrane drying for thicker membranes. It is suggested that membrane thickness should be less than 50 μm in order to mitigate the membrane drying.

## 6. Effect of the Ratio of Thickness of Infected Zone as Compared to the Membrane Thickness δ/d (Figure 54-6)

Locality (in the direction of membrane thickness) of the infected zone is tested, in which the fraction of contaminant ion is fixed at either 5% or 10% and $\delta/d$ is changed. Compared with the case where the contaminant ion distributes uniformly across the membrane ($\delta/d = 0$), local contamination at the cathode side ($\delta/d > 0.8$) gives lower water content and larger membrane resistance overpotential $\eta$(mem). This trend is larger for the cathode side than for the anode side [85]. This result indicates the serious problem of membrane contamination, because contaminants would accumulate more at the cathode side through the air stream.

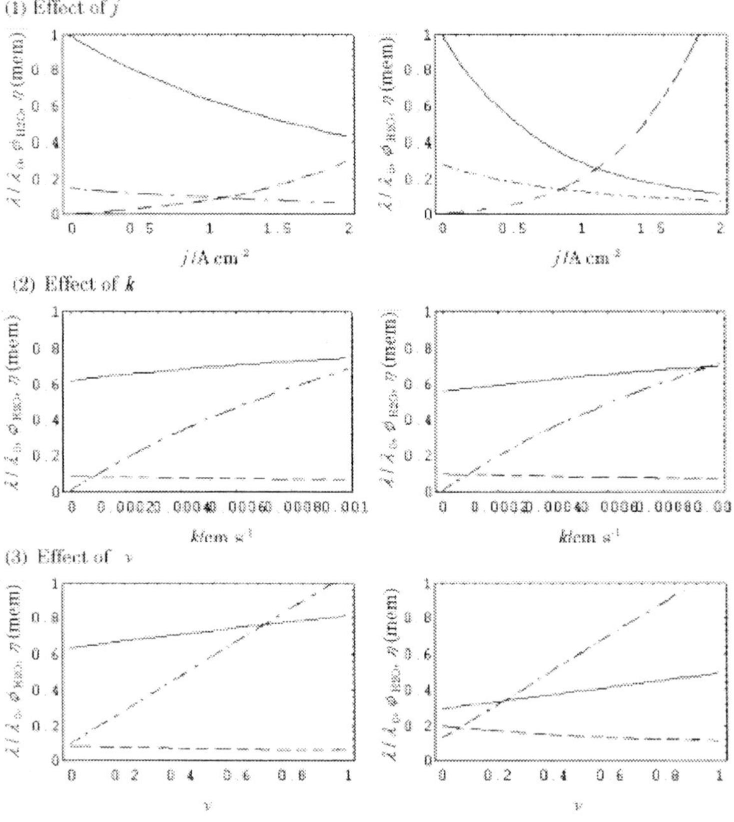

Figure 54. Continued on next page.

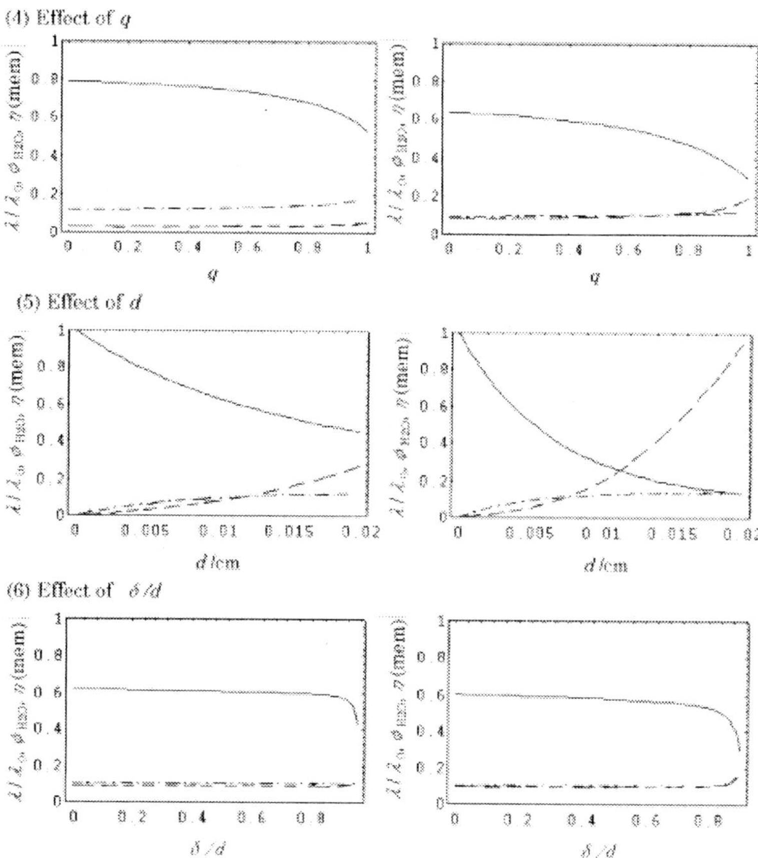


Figure 54. Effect of basic parameters on membrane performances. —: membrane water content as normalized by the initial level $\lambda/\lambda_0$, $-\cdot-\cdot$: net water flux $\varphi_{H2O}$ and ----: membrane resistance overpotential $\eta$(mem)/V as a function of current density $j$ /A cm$^{-2}$. $T$=80°C, $p$ =1 atm, $d$ =100μm, $k$ = 0.0001 cm s$^{-1}$, $v$=0, δ/$d$ = 0.9. (1) Effect of $j$ /A cm$^{-2}$, left hand side: no contamination, right hand side: 10% contamination by Na$^+$. (2) Effect of $k$, left hand side: no contamination, right hand side: 10% contamination by Na$^+$. (3) Effect of $v$, left hand side: no contamination, right hand side: 10% contamination by Na$^+$. (4) Effect of $q$, left hand side: $j$ =0.5 A cm$^{-2}$, right hand side: $j$ =1.0 A cm$^{-2}$. (5) Effect of $d$, left hand side: no contamination, right hand side: 10% contamination by Na$^+$. (6) Effect of δ/$d$, left hand side: 5% contamination by Na$^+$, right hand side: 10% contamination by Na$^+$ (with permission form Elsevier Science).

Among the parameters tested, current density and the membrane thickness show major significance. The present results agree well qualitatively or semi-quantitatively with those obtained experimentally on test fuel cells, and in this sense the method appears as a superior tool considering its ease in calculation, especially when the effect of impurity comes into concern. The importance of the problems of fuel cell degradation by impurity ions is now becoming the issue, and further works are needed on both experimental and simulation studies [81,86].

# CONCLUSION

Transport parameters in perfluorosulfonated ionomer (PFSI) membranes are measured by electrochemical, PGSE-NMR and other methods, and are discussed especially referring to Gierke's cluster network model as polymer microstructure. Ion and water transport occurs in the limited paths of hydrophilic domains of the polymer. The cation conductivity and water permeability in the membranes are affected by the ionic cluster size. The channel cross section (~1 nm diameter) of connecting pores between ionic clusters appears to be the factor that determines the mobility of species inside the polymer. The number of water molecules accompanying cation (water drag) in the membrane increases as the hydration energy of cations increases.

Several states of water exist inside the membrane: semi-free water and bound water are observed by DSC measurements (freezing water and non-freezing water). Semi-free water filled in the ionic channel can be assumed to be those "pushed" by moving hydrated cations through cation migration. Bound water is the hydrated water to cations and sulfonic acid groups, or those in the interphase region between hydrophilic and hydrophobic regions. The number of freezing water per sulfonic acid groups depends on counter cations, and also depends on $EW$, while that of non-freezing water is $EW$ independent but polymer structure dependent. Although FT-IR can not discriminate semi-free water and bound water in the membranes, tendencies are that membranes exchanged with more hydrophilic cations show larger extent of hydrogen bonding and more water signals. Hydrophilic cations with smaller ionic radii would polarize S-O bond more strongly than less hydrophilic cations.

Binary cation systems give much information concerning the molecular interactions in the polymer. Less hydrophilic cations show higher affinity to sulfonic acid groups and the interaction energy between cations and sulfonic acid groups is higher for less hydrophilic cations. Cation-cation interactions are mitigated by shielding effect of water molecules or by sulfonic acid groups. Mobility of $H^+$ in mixed cation systems increases or decreases by the presence of the other cations in different ways. One reason may be the self-exchange rate of water molecules around the cations: in the presence of $Cu^{2+}$ where self-exchange rate of water is high, $H^+$ mobility is accelerated.

From diffusion coefficients and activation energies measured by PGSE-NMR, two mechanisms are proposed for cation movement in the membrane: Grotthuss mechanism and vehicle mechanism, depending on cation species. $Li^+$, $Na^+$ etc. are moving by vehicle mechanism but $H^+$ is mainly moving by Grotthuss mechanism. From the results of PFSI membranes with different *EW* values, it is concluded that the membrane ionic conductivity is determined mainly by $H^+$ mobility rather than by the concentration of sulfonic acid groups. The presence of alcohols significantly lowers the $H^+$ mobility. The difference in the movement of $H^+$ (Grotthuss) and alcohols (vehicle) becomes more apparent for membranes with lower *EW* and smaller alcohols.

Since the cation mobility is a function of the water content in the membrane, it is closely connected with hydrophilic domains of the polymer. The mobility of cations is determined by two major factors: one is the interaction between cations and sulfonic acid groups and the other is the amount of residing water in the ionic channel. These points may be the key to the design of new polymer electrolyte membranes.

A new methodology is developed to cope with the problem of performance degradation of membranes in PEFC by impurity ions. The effects of membrane contamination in operation conditions of fuel cells are analyzed and discussed systematically. By assuming an infected zone of finite thickness, the problem of contamination is handled in an explicit way. The contaminant ion distribution in the membrane affects differently the membrane performance, and at fixed amount of impurities, gives the most serious effect on the degradation of fuel cells, when impurities distribute locally at the cathode | membrane interface.

# Acknowledgments

The authors wish to acknowledge all the people who contributed to this work. Before this project started, T. O. received plentiful suggestions and supports and also warm personal relations with Professor Tormod Førland (late), Professor Katrine Seip Førland (late) and Professor Signe Kjelstrup of Norwegian University of Science and Technology (NTNU) in Norway, when he stayed there in 1990-1991. Together with Professor Signe Kjelstrup, this project had continued in the framework of the Japan-Norway collaboration. During this collaboration, Dr. Lennart Olav Jerdal, Dr. Kenneth Friestad, Mr. Oddvar Gorseth visited Tsukuba and devoted to this project. He is also grateful to Dr. Magnar Ottøy and Dr. Steffen Møller-Holst of NTNU, who gave him a lot of support and friendship during the course of this project. Professor Dick Bedeaux and Dr. Peter A. Cirkel of Leiden University in the Netherlands are greatly acknowledged for their collaboration in the aggregation process of ionomers.

T. O. expresses his hearty thanks to Professor Yasuhiko Ito and Dr. Gang Xie of Kyoto University, who contributed a lot in the difficult time of this project. He acknowledges very much collaboration and helpful discussions with Professor Isao Sekine, Professor Makoto Yuasa and Associate Professor Jun Kuwano of Tokyo University of Science. Thanks are also due to Mr. Norito Nakamura, Dr. Yusuke Ayato, Mr. Hiroki Satou, Mr. Mitsuhiro Okuno, Ms. Naoko Arimura and Mr. Shinya Ikesaka of Tokyo University of Science, who worked with him on the mechanism of membrane transport phenomena.

# REFERENCES


[1] Yeo, R.S. (1982) Chap. 18 Applications of perfluorosulfonated polymer membranes in fuel cells, electrolyzers, and load leveling devices. In: Ref. [10], pp. 453-473.

[2] Heitner-Wirguin, C. (1996) Recent advances in perfluorinated ionomer membranes: structure, properties and applications. *J. Memb. Sci.*, 120, 1-33.

[3] Yeager, H.L.; Gronowski, A.A. (1997) Chap. 8 Membrane Applications. In: Ref. [13], pp. 333-364.

[4] Srinivasan, S.; Manko, D.J.; Koch, H.; Enayetullah, M.A.; Appleby, A.J. (1990) Recent advances in solid polymer electrolyte fuel cell technology with low platinum loading electrodes. *J. Power Sources*, 29, 367-387.

[5] Kreuer, K.D. (1997) On the development of proton conducting materials for technological applications. *Solid State Ionics*, 97, 1-15.

[6] Alberti, G.; Casciola, M. (2001) Solid state protonic conductors, present main applications and future prospects. *Solid State Ionics*, 145, 3-16.

[7] Doyle, M.; Rajendran, G. Chap. 30 Perfluorinated membranes. In: Vielstich, W.; Lamm, A.; Gasteiger, H.A. editors. *Handbook of Fuel Cells: Fundamentals Technology and Applications.* Chichester: Wiley; 2003; Vol. 3; 351-395.

[8] Gierke, T.D.; Hsu, W.Y. (1982) Chap. 13 The cluster-network model of ion clustering in perfluorosulfonated membranes. In: Ref. [10], pp. 283-307.

[9] Zawodzinski, Jr., T.A.; Springer, T.E.; Davey, J.; Jestel, R.; Lopez, C. Valerio, J.; Gottesfeld, S. (1993) A comparative study of water uptake by and transport through ionomeric fuel cell membranes. *J. Electrochem. Soc.*, 140, 1981-1985.

[10] Eisenberg, A.; Yeager, H.L. editors. *Perfluorinated Ionomer Membranes*, ACS Symposium Series 180. Washington D.C.: American Chemical Society; 1982.

[11] Mauritz, K.A.; Hopfinger, A.J. Chap. 6 Structural Properties of Membrane Ionomers. In: Bockris, J.O'M.; Conway B.E.; White, R.E. editors. *Modern Aspects of Electrochemistry*. New York: Plenum; 1982; No. 14; 425-508.

[12] Pineri, M.; Eisenberg, A. editors. *Structure and Properties of Ionomers*. Dordrecht: NATO ASI Series C, D. Reidel Pub. Co.; 1986.

[13] Tant, M.R.; Mauritz, K.A.; Wilkes, G.L. editors. *Ionomers: Synthesis, Structure, Properties and Applications*. London: Blackie Academic and Professional; 1997.

[14] Mauritz, K.; Moore, R.B. (2004). State of Understanding of Nafion. *Chem. Rev.* 104, 4535-4585.

[15] Kreuer, K.-D.; Paddison, S.J.; Spohr, E.; Schuster, M. (2004) Transport in Proton Conductors for Fuel-Cell Applications: Simulations, Elementary Reactions, ad Phenomenology. *Chem. Rev.* 104, 4637-4678.

[16] Harland, C.E. *Ion Exchange: Theory and Practice*. Cambridge: Royal Society of Chemistry; 1994.

[17] Moore, R.B.; Martin, C.R. (1989) Morphology and chemical properties of the Dow perfluorosulfonate ionomers. *Macromolecules*, 22, 3594-3599.

[18] Falk, M. (1980) An infrared study of water in perfluorosulfonate (Nafion) membranes. *Can. J. Chem.*, 58, 1495-1501.

[19] Lowry, S.R.; Mauritz, K.A. (1980) *J. Am. Chem. Soc.*, 102, 4665-4667.

[20] Laporta, M.; Pegoraro, M.; Zanderighi, L. (1999) Perfluorosulfonated membrane (Nafion): FT-IR study of the state of water wit increasing humidity. *Phys. Chem. Chem. Phys.*, 1, 4619-4628.

[21] Eisenberg, A. (1970) Clustering of Ions in Organic Polymers. A Theoretical Approach. *Macromolecules*, 3, 147-154.

[22] Mauritz, K.A.; Rogers, C.E. (1985) A water sorption isotherm model for ionomer membranes with cluster morphologies. *Macromolecules*, 18, 483-491.

[23] Gierke, T.D.; Munn, G.E.; Wilson, F.C. (1981) The morphology in Nafion perfluorinated membrane products, as determined by wide- and small-angle X-ray studies. *J. Polymer Sci., Polym. Phys.*, 19, 1687-1704.

[24] Rubatat, L.; Rollet, A.L.; Gebel, G.; Diat, O. (2002) Evidence of elongated aggregates in Nafion. *Macromolecules*, 35, 4050-4055.

[25] Yeo, S.C.; Eisenberg, A. (1977) Physical properties and supermolecular structures of perfluorinated in-containing (Nafion) polymers. *J. Appl. Polym. Sci.*, 21, 875-898.

[26] Hsu, W.Y.; Gierke, T.D. (1982) Elastic theory for ionic clustering in perfluorinated ionomers. *Macromolecules*, 15, 101-105.

[27] Hsu, W.Y.; Gierke, T.D. (1983) Ion transport and clustering in Nafion perfluorinated membranes. *J. Membrane Sci.*, 13, 307-326.

[28] Yeager, H.L.; Steck, A. (1981) Cation and water diffusion in Nafion ion exchange membranes: influence of polymer structure. *J. Electrochem. Soc.*, 128, 1880-1884.

[29] Mackie, J.S.; Mears, P. (1955) The diffusion of electrolytes in a cation-exchange resin membrane. *Proc. Roy. Soc. Ser. A*, 232, 498-509.

[30] Fernández-Prini, R.; Philipp, M. (1976) Tracer diffusion coefficient of counterions in homo- and heteroionic poly(styrenesulfonate) resins. *J. Phys. Chem.*, 80, 2041-2046.

[31] Yasuda, H.; Lamaze, C.E.; Ikenberry, L.D. (1968) Permeability of solutes through hydrated polymer membranes. Part I. Diffusion of sodium chloride. *Makromol. Chemie*, 118, 19-35.

[32] Verbrugge, M.W.; Hill, R.F. (1990) Transport phenomena in perfluorosulfonic acid membranes during the passage of current. *J. Electrochem. Soc.*, 137, 1131-1138.

[33] Yang, Y.; Pintauro, P.N. (2000) Multicomponent space-charge transport model for ion-exchange membrnanes. *AIChE J.*, 46, 1177-1190.

[34] Paddison, S.J.; Paul, R.; Zawodzinski, Jr., T.A. (2000) A Statistical Mechanical Model of Proton and Water Transport in a Proton Exchange Membrane. *J. Electrochem. Soc.*, 147, 617-626.

[35] Førland, K.S.; Førland, T.; Kelstrup-Ratkje, S. *Irreversible Thermodynamics: Theory and Applications*. New York: John Wiley and Sons; 1988.

[36] Ottøy, M.; Førland, T.; Kjelstrup-Ratkje, S.; Møller-Holst, S. (1992) Membrane transference numbers from a new emf method. *J. Memb. Sci.*, 74, 1-8.

[37] Okada, T.; Møller-Holst, S.; Gorseth, O.; Kjelstrup, S. (1998) Transport and equilibrium properties of Nafion® membranes with $H^+$ and $Na^+$ ions. *J. Electroanal. Chem.* 442, 137-145.

[38] Okada, T.; Kjelstrup-Ratkje, S.; Hanche-Olsen, H. (1992) Water transport in cation exchange membranes. *J. Memb. Sci.*, 66, 179-192.

[39] Xie, G.; Okada, T. (1996) The state of water in Nafion 117 of various cation forms. *DENKI KGAKU*, 64, 718-726.

[40] Laidler, K.J.; Meiser, J.H. *Physical Chemistry*. Menlo Park: Benjamin/Cummings; 1982.

[41] Okada, T.; Nakamura, N.; Yuasa, M.; Sekine, I. (1997) Ion and water transport characteristics in membranes for polymer electrolyte fuel cells containing $H^+$ and $Ca^{2+}$ cations. *J. Electrochem. Soc.*, 144, 2744-2750.

[42] Okada, T.; Ayato, Y.; Yuasa, M.; Sekine, I. (1999) The effect of impurity cations on the transport characteristics of perfluorinated ionomer membranes. *J. Phys. Chem. B, 103*, 3315-3322.

[43] Førland, K.S.; Okada, T.; Kjelstrup-Ratkje, S. (1993) Molten salt regular mixture theory applied to ion exchange membranes. *J. Electrochem. Soc.*, 140, 634-637.

[44] Xie, G.; Okada, T. (1998) Fourier transform infrared spectroscopy study of fully hydrated Nafion membranes of various cation forms. *Z. Phys. Chem.*, 205, 113-125.

[45] Stejskal, E.O. (1965) Use of Spin Echoes in a Pulsed Magnetic-Field Gradient to Study Anisotropic, Restricted Diffusion and Flow. *J. Chem. Phys.*, 43, 3597-3603.

[46] Price, W.S.; Ide, H.; Arata, Y. (2003) Solution Dynamics in Aqueous Monohydric Alcohol Systems. *J. Phys. Chem. A*, 107, 4784-4789.

[47] Weingärtner, H. (1982) Self Diffusion in Liquid Water. A Reassessment. *Z. Phys. Chem. NF (Leipzig)*, 132, 129-149.

[48] Hayamizu, K.; Price, W. S. (2004) A new type of sample tube for reducing convection effects in PGSE-NMR measurements of self-diffusion coefficients of liquid samples. *J. Magn. Reson.*, 167, 328-333.

[49] Aihara, Y.; Sugimoto, K.; Price, W.S.; Hayamizu, K. (2000) Ionic conduction and self-diffusion near infinitesimal concentration in lithium salt-organic solvent electrolytes. *J. Chem. Phys.*, 113, 1981-1991.

[50] Hayamizu, K.; Aihara, Y.; Price, W.S. (2000) Correlating the NMR self-diffusion and relaxation measurements with ionic conductivity in polymer electrolytes composed of cross-linked poly(ethylene oxide-propylene oxide) doped with $LiN(SO_2CF_3)_2$. *J. Chem. Phys.*, 113, 4785 -4793.

[51] Qiao, J.; Hamaya, AT.; Okada, T. (2005) New highly proton conductive polymer membranes poly(vinyl alcohol)-2-acrylamido-2-methyl-1-propanesulfonic acid (PVA-PAMPS). *J. Mater. Chem.*, 15, 4414-4423.

[52] Porat, Z.; Fryer, J.R.; Huxham, M.; Rubinstein, I. (1995) Electron microscopy investigation of the microstructure of Nafion films. *J. Phys. Chem.*, 99, 4667-4671.

[53] Xie, G.; Okada, T. (1996) Characteristics of water transport in relation to microscopic structure in Nafion membranes. *J. Chem. Soc., Faraday Trans.*, 92, 663-669.

[54] Cirkel, P.A.; Okada, T. (1999) Equilibrium aggregation in perfluorinated ionomer solutions. *Macromolecules*, 32, 531-533.

[55] Cirkel, P.A.; Okada, T. (2000) A comparison of mechanical and electrical percolation during the gelling of Nafion solutions. *Macromolecules*, 33, 4921-4925.

[56] Szajdzinska-Pietek, E.; Schlick, S.; Plonka, (1994) A. Self-Assembling of Perfluorinated Polymeric Surfactants in Nonaqueous Solvents. Electron Spin Resonance Spectra of Nitroxide Spin Probes in Nafion Solutions and Swollen Membranes. *Langmuir*, 10, 2188-2196.

[57] Okada, T.; Xie, G.; Gorseth, O.; Kjelstrup, S.; Nakamura, N.; Arimura, T. (1998) Ion and water transport characteristics of Nafion membranes as electrolytes. *Electrochim. Acta*, 43, 3741-3747.

[58] Xie, G.; Okada, T. (1995) Water transoport behavior in Nafion 117 membranes. *J. Electrochem. Soc.*, 142, 3057-3062.

[59] Breslau, B.R.; Miller, I.F. (1971) A Hydrodynamic Model for Electroosmosis. *Ind. Eng. Chem. Fund.*, 10, 554-565.

[60] Xie, G.; Okada, T. (1996) Pumping effects in water movement accompanying cation transport across Nafion 117 membranes. *Electrochim. Acta*, 41, 1569-1571.

[61] Steck, A.; Yeager, H.L. (1980) Water sorption and cation-exchange selectivity of a perfluorosulfonate ion-exchange polymer. *Anal. Chem.*, 52, 1215-1218.

[62] Yoshida, H.; Miura, Y. (1992) Behavior of water in perfluorinated ionomer membranes containing various monovalent cations. *J. Memb. Sci.*, 68, 1-10.

[63] Zawodzinski, Jr., T. A.; Neeman, M.; Sillerud, L. O.; Gottesfeld, S. (1991) Determination of water diffusion coefficients in perfluorosulfonated ionomeric membranes. *J. Phys. Chem.*, 95, 6040-6044.

[64] Zawodzinski, Jr., T. A.; Derouin, C.; Radzinski, S.; Sherman, R. J.; Smith, V. T.; Springer, T. E.; Gottesfeld, S. (1993) Water uptake by and transport through Nafion® 117 membranes. *J. Electrochem. Soc.*, 140, 1041-1047.

[65] Ren, X.; Springer, T. E.; Zawodzinski, T. E.; Gottesfeld, S. (2000) Methanol transport through Nafion membranes electro-osmotic drag effects on potential step measurements. *J. Electrochem. Soc.*, 147, 466-474.

[66] Saito, M.; Arimura, N.; Hayamizu, K.; Okada, T. (2004) Mechanisms of ion and water transport in perfluorosulfonated ionomer membranes for fuel cells. *J. Phys. Chem. B*, 108, 16064-16070.

[67] Thampan, T.; Malhotra, S.; Tang, H.; Datta, R. (2000) Modeling of conductive transport in proton-exchange membranes for fuel cells. *J. Electrochem. Soc.*, 147, 3242-3250.

[68] Saito, M.; Hayamizu, K.; Okada, T. (2005) Temperature dependence of ion and water transport in perfluorinated ionomer membranes for fuel cells. *J. Phys. Chem. B*, 109, 3112-3119.

[69] Cappadonia, M.; Erning, J. W.; Stimming, U. (1994) Proton conduction of Nafion® 117 membrane between 140 K and room temperature. *J. Electroanal. Chem.*, 376, 189-193.

[70] Franks, F., editor. *Water - A Comprehensive Treatise*. New York: Plenum Press; 1975.

[71] Okada, T.; Satou, H.; Okuno, M.; Yuasa, M. (2002) Ion and water transport characteristics of perfluorosulfonated ionomer membranes with $H^+$ and alkali metal cations. *J. Phys. Chem. B*, 106, 1267-1273.

[72] Okada, T.; Arimura, N.; Satou, H.; Yuasa, M.; Kikuchi, T. (2005) Membrane transport characteristics of binary cation systems with $Li^+$ and alkali metal cations in perfluorosulfonated ionomer. *Electrochim. Acta*, 50, 3569-3575.

[73] Holt, T.; Førland, T.; Kelstrup-Ratkje, S. (1985) Cation exchange membranes as solid solutions. *J. Memb. Sci.*, 25, 133-151.

[74] Frank, H.S.; Wen, W.Y. (1957) Structural aspects of ion-solvent interaction in aqueous solutions: a suggested picture of water structure. *Disc. Faraday Soc.*, 24, 133-140.

[75] Wasmus, S; Küver, A. (1999) Methanol oxidation and direct methanol fuel cells: a selective review. *J. Electroanal. Chem.*, 461, 14-31.

[76] Lamy, C.; Léger, J.-M.; Srinivasan, S. Chap 3 Direct Methanol Fuel Cells: From a Twentieth Century Electrochemist's Dream to a Twenty-first Century Emerging Technology. In: Bockris, J.O'M.; Conway, B.E.; White, R.E., editors. *Modern Aspects of Electrochemistry*. New York: Kluwer Academic; 2001; Vol. 34; 53-118.

[77] Saito, M.; Ikesaka, S.; Kuwano, J.; Qiao, J.; Tsuzuki, S.; Hayamizu, K.; Okada, T. (2007) Mechanisms of proton transport in alcohol-penetrated perfluorosulfonated ionomer membranes for fuel cells. *Solid State Ionics*, 178, 539-545.

[78] Saito, M.; Hayamizu, K.; Okada, T. (2006) Alcohol and proton transport in perfluorinated ionomer membranes for fuel cells. *J. Phys. Chem. B*, 110, 24410-24417.

[79] Kreuer, K.D. Chap. 33 Hydrocarbon membranes. In: Vielstich, W.; Lamm, A.; Gasteiger, H.A. editors. *Handbook of Fuel Cells: Fundamentals Technology and Applications*. Chichester: Wiley; 2003; Vol. 3; 420-435.

[80] Morf, W.E. *The Principles of Ion-selective electrodes and of Membrane Transport*. Amsterdam: Elsevier; 1981.

[81] Okada, T. (2001) Modeling polymer electrolyte membrane fuel cell performances. *J. New Mater. Electrochem. Systems*, 4, 209-220.

[82] Okada, T.; Xie, G.; Tanabe, Y. (1996) Theory of water management at the anode side of polymer electrolyte fuel cell membranes. *J. Electroanal. Chem.*, 413, 49-65.

[83] Okada, T.; Xie, G.; Meeg, M. (1998) Simulation for water management in membranes for polymer electrolyte fuel cells. *Electrochim. Acta*, 43, 2141-2155.

[84] Okada, T. (1999) Theory for water management in membranes for polymer electrolyte fuel cells. Part 1. The effect of impurity ions at the anode side on the membrane performances. *J. Electroanal. Chem.*, 465, 1-17.

[85] Okada, T. (1999) Theory for water management in membranes for polymer electrolyte fuel cells. Part 2. The effect of impurity ions at the cathode side on the membrane performances. *J. Electroanal. Chem.*, 465, 18-29.

# INDEX

## D

DuPont, 6
durability, 82
duration, 31

## E

earth, 30, 43, 44, 45, 47, 48, 52
Einstein, 33, 34, 35
elastic deformation, 8
electric conductivity, 20, 39
electric current, 23
electric field, 5
electric potential, 16
electrical, 39, 105
electricity, 20
electrochemical, 1, 3, 11, 17, 19, 21, 23, 24, 25, 34, 97
electrochemistry, 34
electrodes, 21, 22, 26, 101, 106
electrolysis, 1, 6, 8, 12
electrolytes, 1, 19, 34, 56, 70, 80, 84, 86, 91, 98, 101, 104, 107
electromotive force, 21, 22, 23
electron, 7, 28, 37, 38
electronic, 34
electrostatic, 7, 8, 44, 79, 82
electrostatic force, 8, 79, 82
emission, 28
energy, 1, 8, 12, 14, 17, 28, 38, 66, 67, 76, 97, 98
entropy, 8, 49, 67
environment, 1, 28
equilibrium, 21, 28, 29, 39, 49, 65, 66, 68, 103
equilibrium state, 49, 66
ethanol, 35, 39
ethylene, 104
evaporation, 39
evolution, 39
exchange rate, 62, 98
exclusion, 44, 45, 47, 50
exothermic, 58
extraction, 1

## F

fabrication, 5, 82
failure, 47
films, 6, 38, 50, 51, 104
flooding, 85
flow, 17, 23, 25, 47, 85
fluorescence, 28
fluorinated, 49
food, 6
Fourier, 30, 104
Fourier transform infrared spectroscopy, 104
free energy, 17
free volume, 14
freedom, 41
freezing, 30, 49, 50, 51, 57, 58, 59, 61, 97
friction, 79, 82
friendship, 99
FT-IR, 30, 50, 51, 97, 102
fuel, 1, 6, 70, 72, 82, 84, 85, 95, 98, 101, 104, 105, 106, 107

## G

gas, 25, 85, 89
gas diffusion, 85
Gaussian, 8
Gibbs, 21, 66
glass, 7, 35
gravity, 33
groups, 1, 5, 6, 7, 8, 10, 17, 19, 27, 29, 30, 37, 39, 43, 45, 49, 50, 52, 53, 65, 67, 69, 70, 73, 79, 80, 82, 97, 98

## H

$H_2$, 19
Hamiltonian, 17
height, 31, 33
high resolution, 31
homogeneous, 6, 33, 35, 86
host, 80
humidity, 102
hydrate, 45

## N





## O



## P

## T





## U



## V